NATURE ANIMATED

VOLUME II

THE UNIVERSITY OF WESTERN ONTARIO SERIES IN PHILOSOPHY OF SCIENCE

A SERIES OF BOOKS

IN PHILOSOPHY OF SCIENCE, METHODOLOGY,

EPISTEMOLOGY, LOGIC, HISTORY OF SCIENCE,

AND RELATED FIELDS

VOLUME 21

NATURE ANIMATED

Historical and Philosophical Case Studies in Greek Medicine,
Nineteenth-Century and Recent Biology,
Psychiatry, and Psychoanalysis

Papers Deriving from the Third International Conference on the History and Philosophy of Science, Montreal, Canada, 1980

VOLUME II

Edited by

MICHAEL RUSE
Departments of History and Philosophy, University of Guelph

D. REIDEL PUBLISHING COMPANY

DORDRECHT: HOLLAND / BOSTON: U.S.A.

LONDON: ENGLAND

Library of Congress Cataloging in Publication Data

CIP

International Conference on the History and Philosophy
of Science (3rd : 1980 : Montréal, Québec)
Nature animated.

(The University of Western Ontario series in
philosophy of science ; v. 21)
Includes bibliographical references and indexes.
1. Science—Philosophy—Congresses. 2. Science—
History—Congresses. I. Ruse, Michael. II. Title.
III. Series.
Q174.I563 1980 501 82–9013
ISBN-13: 978-94-009-6962-9 e-ISBN-13:978-94-009-6960-5
DOI: 10.1007/978-94-009-6960-5

Published by D. Reidel Publishing Company,
P.O. Box 17, 3300 AA Dordrecht, Holland

Sold and distributed in the U.S.A. and Canada
by Kluwer Boston Inc.,
190 Old Derby Street, Hingham, MA 02043, U.S.A.

In all other countries, sold and distributed
by Kluwer Academic Publishers Group
P.O. Box 322, 3300 AH Dordrecht, Holland

D. Reidel Publishing Company is a member of the Kluwer Group

TABLE OF CONTENTS

VOLUME II

TABLE OF CONTENTS

VOLUME I

PART IV

PART V

PREFACE

These remarks preface two volumes consisting of the proceedings of the Third International Conference on the History and Philosophy of Science of the International Union of History and Philosophy of Science. The conference was held under the auspices of the Union, The Social Sciences and Humanities Research Council of Canada, and the Canadian Society for History and Philosophy of Science. The meetings took place in Montreal, Canada, 25–29 August 1980, with Concordia University as host institution.

The program of the conference was arranged by a Joint Commission of the International Union of History and Philosophy of Science consisting of Robert E. Butts (Canada), John Murdoch (U.S.A.), Vladimir Kirsanov (U.S.S.R.), and Paul Weingartner (Austria). The Local Arrangements Committee consisted of Stanley G. French, Chair (Concordia), Michel Paradis, treasurer (McGill), François Duchesneau (Université de Montréal), Robert Nadeau (Université du Québec à Montréal), and William Shea (McGill University). Both committees are indebted to Dr. G. R. Paterson, then President of the Canadian Society for History and Philosophy of Science, who shared his expertise in many ways. Dr. French and his staff worked diligently and efficiently on behalf of all participants. The city of Montreal was, as always, the subtle mixture of extravagance, charm, warmth and excitement that retains her status as the jewel of Canadian cities.

The funding of major international conferences is always a problem. This conference was exceptional in that financial support came forward from many sources. Contributions from the Division of History of Science and the Division of Logic, Methodology and Philosophy of Science of the International Union of History and Philosophy of Science were matched by a seed grant from the Social Sciences and Humanities Research Council of Canada. The seeds bore fruit in the form of grants from Concordia University, McGill University, Université de Montréal, University of Calgary, University of Guelph, University of Lethbridge, The University of Western Ontario, the Hannah Institute for the History of Medicine, and The University of Western Ontario Series in Philosophy of Science. On behalf of the IUHPS, the members of the Joint Commission thank all sponsors for their generous support, and for their cooperative demonstration of the fact that there is, indeed, strength in numbers!

Michael Ruse (ed.), Nature Animated, ix–x.

The conference organizers owe many additional debts which we here gratefully acknowledge. We thank the many people who worked for Dr. French, but especially Stephanie Manuel, who organized most of the details of local arrangements, Shona French, a gracious and helpful lady during a long week, and Susan Hudson, who designed the conference posters and the conference program booklet, always useful items but in this case elevated to a new standard of aesthetic attraction. We thank all of those who were our hosts – for drinks between sessions, for lunch or dinner; especially we thank William Shea and Mario Bunge for a marvellous luncheon at McGill, Dr. and Mrs. Robert V. V. Nicholls, host and hostess at a glittering reception in the storied Ritz-Carlton Hotel, and the administration of Concordia University, for a second welcoming reception in the Faculty Club of Concordia.

Comparison of the tables of contents of the two conference volumes with the final program of the conference will reveal that not all papers delivered at the meetings were made available for publication. In some cases, material appearing in the proceedings is largely new (Albury, Moore), or very extensively revised (Duchesneau, Lennon, Okruhlik). The conference contribution of Jonathan Hodge grew to monograph length in revision; it is our hope that it might appear separately in a short time. The papers by Joly and Clavelin were delivered at the conference in French. We thank Vida Bruce for the translation of Joly's paper, and Violaine Arès and Stella-Marie Baza for the translation of Clavelin's paper. In both cases the translations were corrected and approved by the authors.

We are especially grateful to Michael Ruse and William Shea for their willing acceptance of the chores of editing the volumes, and for writing the special introductions to each of the volumes. Mrs. Nel Jones, WOS editor at Reidel, was patient as usual, and as usual enormously helpful. We are very grateful. Judith York, editorial assistant in the WOS office at The University of Western Ontario, edited the difficult papers and organized the final typescripts. We owe her many thanks. She was a cog in the wheels at Western, as was Pat Orphan, who did all of the thousand little things connected with getting the project launched, things usually ignored or forgotten. We remember and take notice, with sincere thanks.

On behalf of editors and organizers,

October, 1981 ROBERT E. BUTTS

PROGRAM

THIRD INTERNATIONAL CONFERENCE ON HISTORY AND PHILOSOPHY OF SCIENCE, CONCORDIA UNIVERSITY, MONTREAL, CANADA, 25–29 AUGUST, 1980

CONCEPTUAL AND EMPIRICAL PROBLEMS IN ANCIENT GREEK MEDICINE AND NATURAL PHILOSOPHY

Paul Potter (Canada), 'Hippocratic Nosology: the Identification and Classification of Diseases'.
John Wright (Canada), Comments on Potter's paper.
Robert Joly (Belgium), 'Hippocrate et l'Ecole de Cos: entre le mythe et l'hypercritique'.
Jaap Mansfeld (The Netherlands), Comments on Joly's paper.

In the Chair: Maurice Lebel (Canada).

HISTORY AND PHILOSOPHY OF PSYCHIATRY

William Bynum (U.K.), 'Themes in British Psychiatry, J. C. Pritchard (1785–1848) to Henry Maudsley (1835–1918)'.
Stefano Poggi (Italy), Comments on Bynum's paper.
Morris Eagle (Canada), 'Anatomy of the Self in Psychoanalytic Theory'.
Michael Moore (U.S.A.), Comments on Eagle's paper.
David Gruender (U.S.A.), Comments on Eagle's paper.

In the Chair: Eva Lester (Canada).

MATHEMATICS IN THE DEVELOPMENT OF SCIENTIFIC THEORIES

Paul Weingartner (Austria), 'The Ideal of the Mathematization of All Sciences and of 'More geometrico' in Descartes and Leibniz'.
François Duchesneau (Canada), Comments on Weingartner's paper.
Thomas Lennon (Canada), Comments on Weingartner's paper.
V. Vizgin (U.S.S.R.), 'The Genesis of the Tensor Geometrical Conception of Gravitation'.

In the Chair: Joseph Pitt (U.S.A.).

Michael Ruse (ed.), Nature Animated, xi–xiii.

HISTORY AND PHILOSOPHY OF 19TH CENTURY BIOLOGY

Camille Limoges (Canada), 'L'économie politique d'une tendance hégémonique en histoire naturelle: le cuviérisme en France au XIXe siècle'.
W. R. Albury (Australia), Comments on Limoges' paper.
M. Jaroshevsky (U.S.S.R.), 'The Problem of Voluntary Motion Determination in Russian Psycho-Physiology'.
Jonathan Hodge (U.K.), 'Darwin on Natural Selection: his Methods and his Methodology'.
Michael Ruse (Canada), Comments on Hodge's paper.
John Beatty (U.S.A.), 'What's in a Word? The Problem of Coming to Terms in the Darwinian Revolution'.
David Hull (U.S.A.), Comments on Beatty's paper.
A. Lindenmayer (The Netherlands), Comments on Beatty's paper.

In the Chair: Mario Bunge (Canada).

MATHEMATICS AND PHYSICS IN THE 17TH AND 18TH CENTURIES

Maurice Clavelin (France), 'Aspects conceptuels et techniques de la géometrisation Galiléene du mouvement de graves'.
William Shea (Canada), Comments on Clavelin's paper.
J. E. McGuire (U.S.A.), 'Space, Geometry and Infinity: Newton and Descartes on the Indefiniteness of Extension'.
John North (The Netherlands), 'Finite and Otherwise. Some Seventeenth Century Views'.
Peter Harman (U.K.), 'Force and Inertia: Euler and Kant's *Metaphysical Foundations of Natural Science*'.
Kathleen Okruhlik (Canada), Comments on Harman's paper.
V. Kirsanov (U.S.S.R.), 'Non-Mechanistic Ideas in Physics and Philosophy: from Newton to Kant'.
V. Kartsev (U.S.S.R.), 'Hypothetical Experiments of Academician Petrov and Electrical Experiments of the XVIII Century'.

In the Chair: Roger Angel (Canada).

Communications not read in sessions:

MATHEMATICS IN THE DEVELOPMENT OF SCIENTIFIC THEORIES

Karel Berka (Czechoslovakia), 'The Ideal of Mathematization in B. Bolzano'.

MATHEMATICS AND PHYSICS IN THE 17TH AND 18TH CENTURIES

Ashot Grigorian (U.S.S.R.), 'Measure, Proportion and the Mathematical Structure of Galileo's Mechanics'. (Copied for distribution).

INTRODUCTION

MICHAEL RUSE

THE NEW DUALISM: "RES PHILOSOPHICA" AND "RES HISTORICA"

All I wish to assert is that there exists an enterprise which is taken seriously by everyone in the business where simplicity, confirmation, empirical content are discussed by considering statements of the form (x) (Ax ⟶ Bx) and their relation to statements of the form Aa, Ab, Aa & Ba, and so on and *this* enterprise, I assert, has nothing whatever to do with what goes on in the sciences. Thee is not a single discovery in this field (assuming there have been discoveries) that would enable us to attack important scientific problems in a new way or to better understand the manner in which progress was made in the past. Besides, the enterprise soon got entangled with itself (paradox of confirmation; counterfactuals; grue) so that the main issue is now its own survival and *not* the structure of science. That this struggle for survival is interesting to watch I am the last one to deny. What I do deny is that physics, or biology, or psychology can profit from participating in it. It is much more likely that they will be *retarded*. (Paul Feyerabend, 1970)

The Montreal conference of 1980 brought together philosophers and historians, to work together trying to understand aspects of the sciences: physical, biological, and social. Obviously, the participants were a biased selection, in that the organizers deliberately chose scholars who think that philosophers and historians can profitably work together; but, the group was not that biased. Today, the notion that philosophy and history can mutually and profitably work together excites no great surprise, from either philosophers or historians of science, and there is an increasing number of people who try simultaneously to work in both fields. (I may be wrong, but I have a sneaking suspicion that most of these "dualists" started first in philosophy, and then moved on to history as well, rather than *vice versa*.)

This friendliness between philosophy and history is a relatively new phenomenon. Less than ten years ago, a well-known philosopher of science, caught for his sins (or perhaps, for his colleagues' sins?) in a department of history and philosophy of science, pondered whether the union was an intimate connexion or a marriage of convenience (Giere, 1973). He had little trouble in concluding that it was the latter: there is no true intellectual sympathy between the two disciplines; rather, we unite only because a medium-sized department fares far better in the academic struggle for existence than do two small-sized departments. Conversely, many historians of science had near-contempt for philosophers of science: a contempt which

Michael Ruse (ed.), Nature Animated, 3–25.

was not entirely without justification, when leading philosophers of science openly and brazenly admitted that they falsified the true story of science, if and when such a story did not fit well with the philosophical tale that they were telling.

We have indeed come a long way since those days, as I trust the fascinating papers in this volume, and its companion, amply testify. No doubt, we have much further to go, and our students will look back with amused pity at our fumbling attempts to integrate and exploit the twin subjects of philosophy and history of science. Since the links between grandparents and grandchildren is often much warmer than that between parents and children, we can only hope that our students' students will look a little more benevolently at our efforts!

I suppose that, as editor and writer of the introduction to this collection of contributions to the Montreal conference, contributions which were devoted to the biological and social sciences, I ought to give you the benefits of my distilled wisdom about the true connexions between philosophy and history of science, together with my suggestions for future work and my forecasts of anticipated successes. At least, this is the sort of thing that editors usually do in these kinds of circumstances: a function, no doubt, of the fact that for an editor, editorializing is like a preacher preaching, in that no one is able to answer back. (Of course, you can turn on the pages of this volume very much more readily than you can get up in the middle of a sermon and walk out of church.)

However, I have discovered – to my chagrin and to your relief – that I have nothing very profound to say to you. Certainly, I have no wide-sweeping conclusions to draw, nor fascinating predictions to make. This is perhaps just as well, for philosophers talking about the future invariably get it wrong. I still have colleagues who are predicting the impossibility of molecular biology.

So instead, what I want to do in this introduction is something very much more modest; but, I hope, not without interest and a certain worth. I shall tell you how one person, namely myself, came to do both philosophy and history of science, and how I see that the two fields have interacted profitably, in my own work. You may disagree with me on various points, and you may disagree with my conclusions. This will not worry me at all. This introduction will have succeeded if I can convert some who still doubt that there is indeed a point of useful contact between philosophy and history of science. It will have succeeded beyond all measure, if I can persuade some readers to take up the very problems which have engaged me, and to do a better job with them than I.

PAST AND PRESENT

About twenty years ago I escaped, a refugee from a very unpleasant maths degree programme. I had nothing but the intellectual clothes on my back: I certainly carried no honours in my knapsack. Fortuitously, I came to philosophy, entered in, founded that I liked it, and decided to stay. Obviously, I had to learn the language and the customs, and, in due course of time, to think about applying for citizenship. More prosaically, I had to take courses, read the literature, and take exams, first for an undergraduate degree, and then later for graduate degrees.

I was increasingly drawn towards the philosophy of science. This was partly because of my background, and partly because, in those days, ethics was such an incredibly boring subject. If no other good came out of the Viet Nam war, it surely convinced moral philosophers that there are more important issues than returning library books on time, and more to human thought and behaviour than not committing the naturalistic fallacy.

My PhD thesis was on the philosophy of biology, specifically the nature of evolutionary theory. This was a perfect choice, because the literature on the subject was limited, and much of it was very bad. It was also, coincidentally, a fascinating topic, and I have never regretted my involvement with the biological sciences. I should say, also, that I have received much help and encouragement from biologists themselves. The most distinguished of men have given unstintingly of their time and knowledge. This help has been paralleled by that from philosophers. The names of Ernst Mayr and of David Hull must not go unmentioned in this context.

The point I want to make here is not about the quality of my early work – I suppose it was no better and no worse than that of many others – but about its totally ahistorical nature. My primers had been R. B. Braithwaite's *Scientific Explanation*; Ernest Nagel's *The Structure of Science*; and Israel Scheffler's *Anatomy of Inquiry*. I had perhaps moved beyond "grue" and white swans; but, both my PhD thesis and the book which grew from it, had the same insensitivity towards the past, as had these classic "logical empiricist" texts. Like them, I drew the same hard line between the context of discovery and the context of justification; like them, I then ignored the former; and, like them, it was no wonder that I produced a frozen, snapshot picture of science, where explanations and predictions hover in timeless eternity, rather like disembodied Platonic forms. Lemming-like, with my contemporaries, I learnt to curl my lip slightly at the mention of those philosophers, like Stephen Toulmin and the late Norwood Russell Hanson, who pleaded that there is more to science than its present.

My road to Damascus – my conversion to the worth of the history of science – was two-fold; although, I suspect, not all that unique. First, I was privileged to see Hanson himself in action, just before his untimely death. (It was, in fact, at a conference on Newton, at the University of Western Ontario.) Those who knew Hanson will remember his charisma, and will recall what an impact he could have. If a man who was this bright, this articulate, this attractive, thought that history of science was important, who was I to say otherwise? Second, I read Thomas Kuhn's inspiring book, *The Structure of Scientific Revolutions*. Like everyone else, I was infuriated at Kuhn's relativistic philosophy; and, like everyone else, I was charmed by Kuhn's easy style, and overwhelmed by the mass of historical example that he threw at us.

Cutting a long story short, and ruthlessly suppressing memories of my first excursions into the history of evolutionary biology – I still remember the cries of anguish from that gentle, kind man, John C. Greene, when I proudly sent him my earliest productions in that field – I labored long and hard in the decade following, in the history as well as in the philosophy of Darwinism. I can make no judgements myself about the quality of my work. It is enough for me that I have now redeemed myself in John Greene's eyes. However, candor does force me to admit that my various efforts have not been met with unbounded enthusiasm. I am perversely fondest of one review which begins: "Personally I find this book rather offensive." This, a review of something in the history of science, is balanced by a review (by one of the contributors to this present volume no less) which concludes that the work is 'dangerously misleading'. I hasten to add that this conclusion comes at the end of a scrupulously fair and detailed exposition of what I had said. Perhaps, that makes it all the more devastating.

The point of relevance is that, after my efforts, I felt I could pride myself on being, simultaneously, both a philosopher and a historian of science. I even held appointments in both the philosophy and history departments of my university. (A dubious status, which obligates me to attend the faculty meetings of both departments; but, which does not entitle me to two offices, or to increased secretarial assistance). However, I now see that, until a short time ago, I was curiously schizophrenic. I did philosophy of science. I did history of science. And yet, I never consciously did philosophy and history of science – or, history and philosophy of science. At the level of conscious reflection and intention, I kept the two subjects apart.

A number of facts made me realize this. I attended a number of first-class conferences, where philosophers and historians tried consciously to work together, for the benefit of each and all. Three conferences, in particular, were worthy of note. First, there was a conference sponsored by the University of Pittsburgh, organized by Larry Laudan, on nineteenth-century scientific methodologies (1975); second, there was the first Leonard conference of the University of Reno, Nevada, organized by Tom Nickles, on scientific discovery (1978); and third, there was this more-general Montreal conference, organized by Robert Butts and Stanley French (1980).[1]

Then again, making me realize what a strangely disjointed life I led, I was privileged to go as a visitor to departments where history and philosophy of science are jointly taken seriously, namely Indiana University's History and Philosophy of Science Department, and the University of Western Ontario's Philosophy Department. And finally, perhaps most important of all, a number of younger scholars – taking the importance of the union of history and philosophy of science for granted – have set me thinking about the history/philosophy interface. I recommend to your attention the theses of John Beatty (Indiana, 1978) and James Robert Brown (University of Western Ontario 1981).[2]

These various influences set me to thinking about the relationships between the philosophy and history of science, especially as they had occurred in my own work. So now, let me share my conclusions with you. I do want to emphasize again that my aim here is not to blow my own trumpet or to convince you of my conclusions. I simply want to show you some examples of how I now see that, philosophy and history of science can gain from each other.

THE PHILOSOPHY OF SCIENTIFIC CHANGE

Start with philosophy. In what sense can a knowledge of the history of science benefit a philosopher? My own experience tells me that there are at least two major ways. First, we all now realize – thanks in no small part to people like Kuhn and Toulmin – that there is more to science than the timeless present. There is more to science than is presented in the journals of today; than is given in the rounded, confident perfection of elementary textbooks; and than is presupposed in the exams of undergraduate degree programmes.

Science has a temporal dimension. A scientific theory, or discipline, grows, thrives, and dies, to use a popular organic metaphor. Then another science takes over, and the overall process goes on. The history of science teaches us this: if it does not teach us this, then it teaches us nothing. Darwin did not just sit down one day in his study in 1858, think up the theory of evolution through natural selection, and write the *Origin of Species*. His ideas came slowly, over the years, from many sources; and, when they arrived publicly, they had to battle for supremacy with rival ideas (de Beer, 1963; Eiseley, 1958). It is, therefore, incumbent on a philosopher of science to study this temporal dimension to science. You may not want to; but then, no one said you had to be a philosopher of science! A philosopher of science is someone who tries to take a look at science, from outside as it were. He (and increasingly and thankfully, she) tries to understand what science is, what makes it work, what makes it change, what makes it good and what makes it bad, and what makes it go. Since science does exist in time as well as space, this means that the philosopher as philosopher simply as to look at the temporal dimension. If nothing else, the history of science has certainly convinced me that a work like Scheffler's *Anatomy of Inquiry* is only doing half of its job.

If only for the sake of argument, let us grant now that the history of science teaches us about the temporal dimension to science, also draws the philosopher's attention to related problems she/he must face and try to solve. What is the nature of scientific change? For instance, is it evolutionary or revolutionary? And, what makes for scientific change? Is it a rational process, or is it governed by all the dark elements to which human emotions and actions are subject?

Since the history of science drew our attention to these problems, can we – must we – look to it for help in their solutions? I believe so; although, admittedly, not everyone would agree with me on this. Scheffler himself has written a book (*Science and Subjectivity*) which "disproves" the Kuhnian position on science's history: a book which contains nary a smell of science's history itself. And, there are other works in the same *genre*. Carl Kordig's *The Justification of Scientific Change* springs to mind.

I am not quite sure what to say in response to people like this. To borrow a term which they will certainly not like, we view science from completely different paradigms. Our very approaches to science itself are so different, as almost to preclude any rational discourse. I feel very much as when I, an evolutionist, am asked to debate with Creationists. What can I say to a man who believes that the earth is only 6000 years old? Analogously, what can I say to a man who believes that he can talk meaningfully about scientific change, when he is totally ignorant of the history of science? My main consolation – and I think this a point which has philosophical as well as sociological merit – is that works like Scheffler's and Kordig's today seem to have all the vitality of the dodo: they are extinct. Those of us still concerned about the nature of scientific change turn, without question, to the history of science.

Perhaps it is simplest if I just state my own position flatly. I do not see how one can apply logic and analysis to something one does not know about. One cannot prove that space is Euclidean or non-Euclidean, without looking at space. In like fashion, one cannot prove that science is Kuhnian or non-Kuhnian, without looking at science. It matters not at all how sophisticated or highflalutin one's logical productions may be, if they do not tell us how science works and how scientists behave. Perhaps it was irrational for the anatomist Richard Owen to oppose the *Origin*; perhaps it was not. But, given that everyone recognized Owen as one of the leaders of science of his day, a philosophy of science which has no place for Owen is no true philosophy. And the same goes for all the other actions of scientists, be they sensible or not by modern standards. And how else than through history can you even know that your philosophy must account for the Owens of this world?

In short, one *must* judge one's philosophical conclusions against historical reality, and probably if one has not been sensitive to history throughout one's philosophising, one's conclusions will not fare that well. As Brown argues very convincingly:

> That [philosophical] methodology is best which makes its theoretical reconstructions and normative reconstructions coincide for the greatest number of episodes in the history of science, and which best coheres with other accepted theories. (p. 72)

To this, I would add that if one is going to do the job properly, one must not simply genuflect piously in the direction of the history of science. One must go further, and one must really get to grips with the subject. Good analyses of the nature of theory change will not emerge from reading popular secondary sources. One must become as good a historian as the historian him/ herself. That means one must read primary sources and unpublished documents, as well as pertinent scholarly books and articles. Otherwise, one stands in the gravest danger of reporting, not on science, but on the philosophy of the historian(s) being used. As we shall see, I do not berate the historians for having philosophies. But, philosophers of all people must choose their own philosophies.

Having come this far, you may wonder what particular philosophy of scientific change my historicising had led me to? I am afraid that my answer is rather a limp one, for I must confess to being in a frightful muddle! My main consolation is that, at least, like Socrates I am that much farther ahead in knowing that I do not know. As a former logical empiricist, I was, and still am, attracted to the rationality of science; and, my studies of both the Darwinian and recent geological revolutions have convinced me that the course of science is certainly not totally without sane reason (Ruse, 1979; Ruse 1982a, b). There is some sort of progression towards true knowledge of objective reality – a progression governed by reason and the empirical facts. But, history has also convinced me that there is a strong subjective (mind-given) element to science too, about which I shall have more to say shortly.

However, as I have said, my aim here is not to push my own views, but to show how philosophy needs history. And, this I hope I had done, in a non-limp fashion. My own inabilities as a philosopher are, fortunately, not really pertinent here.

LEARNING FROM THE PAST: DISCOVERY

There is a second way in which the philosopher needs the history of science. It is the way in which we all need history: to throw light on the present. It is so much easier to cast beams out of the eyes of others, than to deal with the motes in our own eyes. It is so much easier to see what people did in other generations, than to see what we do in our own generation.

In other words, if we are prepared to explore the works of other generations, and *if* (a crucial "if"!) we are then prepared to apply our findings to the present, we can find out much about ourselves which we would not normally spot directly. In particular, the history of science – the story of past science – can tell us a very great deal about present science. Logically, perhaps, we do not need the history. As fallible mortals, caught in our own generation, we do. Let me give three examples, from my own experience.

First, the history of science has convinced me, as a philosopher studying the nature of science, of the importance of the context of discovery. I mentioned earlier how one was taught to draw a rigid distinction between the context of discovery and that of justification, and how one was urged to ignore the former.

Typical, for instance, is Carl Hempel's (excellent) little textbook, *The Philosophy of Natural Science*, which has a delightful introduction to science, dealing with Semmelweis's discovery of the cause of childbed fever, but which then says nothing more about discovery.

Indeed, many philosophers, like Braithwaite and Mario Bunge (1968), warn against too close a familiarity with discovery, lest one illegitimately read into the finished science, some element encountered on the road to that science. One will start asking whether the benzene ring is cold-blooded, and if a proper application of Archimedes Principle requires bath salts!

My own study of the Darwinian Revolution has shown me that this attitude is quite wrong – judged philosophically (Ruse, 1980). The aim of the philosopher is to understand the nature of science – why certain elements are as they are, and what the various connections are between them. Now, if science were solely a disinterested reporting of the empirical facts and of the actual relations between them, I agree that the scientist's route to discovery would be irrelevant. And, indeed, I agree fully that science, in respects, is such a disinterested endeavour. There seems, for instance, to be a genuine struggle for existence and a consequent natural selection, in the world. That Darwin discovered them through affirming the analogy of artificial selection, and that Wallace got to them through denying the analogy of artificial selection, is irrelevant to the actuality of the struggle and of selection: those very concepts described and discussed in the *Origin of Species*.

But, of course, there is more to science than this. One hopes to persuade the reader of the validity and importance of one's claims. To this end, one goes beyond the bare reporting of facts, using language, metaphors, and examples. And these, crucial parts of the theory, can be understood only in the context of discovery. Thus, for instance, in the *Origin*, Darwin brings the reader to natural selection, through the self-same analogy of artificial selection. Then, he uses the term "selection" for his mechanism. And, for no other reason than that it reflects a division made by breeders (selection for profit and selection for pleasure), Darwin divides his wild type of selection into natural selection and sexual selection.

Simply put: these various factors just listed are parts of the *Origin*; they reflect the route of discovery; they were certainly argued about in the acceptance of Darwinism; they are therefore of concern to the philosopher. And, I suspect, the same holds for today's science. (Think for a moment over the row about the term 'selfish gene'.) The philosopher ignores discovery at his/her peril.

LEARNING FROM THE PAST: REGULATIVE PRINCIPLES

Second, history has convinced me of the importance in science of what Kantians call 'regulative principles": rules or structures according to which good science must be formed, if indeed it is to be called "science" at all (Korner, 1960). Consider the problem of teleology, something which has wracked and engrossed philosophers for several years now (mainly because they think, quite mistakenly, that one needs to know no biology at all to talk knowledgeably about it). Biologists talk of 'functions', and 'ends', and 'purposes'. Physicists never do. Thus, the biologist can ask about the purpose of the heart, or of the Dimetrodon's sail fin. A physicist would be laughed out of court, if he asked what the purpose of the moon is.

Why is there this difference? The answer, for me, came in studying the Darwinian Revolution (Ruse, 1981). Charles Darwin produced an epoch-making work, taking us from the miraculously interfering Great Designer of William Paley, to a world governed by blind invariant law. And yet, from the point of view of teleology, it made no difference at all! Before Darwin, people said that the eye exists, in order that we may see. After Paley, people said that the eye exists, in order that we may see. Why is this?

The answer is to be found by looking at those who came first, namely the Paleyites. They said that the eye exists, in order that we may see, because *the eye is like a telescope*. In other words, the eye is like an object of human functional design, and thus we think it appropriate to use the teleological language we use of human functional objects. We make objects with ends in view, and hence we use appropriate language. For Paley and followers, it made similar sense to use such language for what were (then) taken to be God's artifacts.

Now, the fact that organic features seem *as though* designed, as though made, continued unchanged after the *Origin* was published. Darwin's proposing a new origin made no difference to the way in which organic features present themselves to the observer. Hence, to Darwinians, it was considered appropriate and natural to go on using the language of teleology. Organic features are still artifactlike, and were considered as such. And here, we have the reason for teleology today. The eye is like a telescope, so we treat it as such. The moon is not like an artifact, and so we do not treat it as such.

But, see what this little history story all entails. Biologists – pre- and post-Darwin – impose upon the organic world a particular way of looking at it. We interpret it, as if it were designed. We think of it in this way, and conceptualize it as such. We do not put such an interpretation on the inorganic world. And, this is all what I mean when I talk of using regulative principles: history has taught me that biologists view their world through the lens of the artifact model or metaphor. Logically, they do not have to. In fact, they do and find it profitable to do so. But, one must see that what is happening here is an imposition upon experience, rather than something being read directly from experience.

Let me quote my favourite philosopher of science on the subject. Appropriately, he was both a pre-Darwinian and one who influenced Darwin:

> Thus we necessarily include, in our Idea of Organization, the notion of an End, a Purpose, a Design; or, to use another phrase which has been peculiarly appropriated in this case, a *Final Cause*. This idea of a Final Cause is an essential condition in order to the pursuing our researches respecting organized bodies.
>
> This Idea of Final Cause is not *deduced* from the phenomena by reasoning, but is *assumed* as the only condition under which we can reason on such subjects at all. (Whewell, 1840, p. 620)

LEARNING FROM THE PAST: IDEOLOGY

Third and finally, showing how history can throw light on the present, let me refer very briefly to the question of values and ideology in science. There is a popular opinion among analytic philosophers, at least there was until very recently, that science is value-free or neutral. Unlike a political polemic or religious sermon, the scientist's own beliefs, fears, hopes, joys, prejudices, do not intrude into his/her science. It is true that science can be used for certain ends – good or bad – but science itself is neutral. The suicide falling from the CN Tower may regret Galileo's laws; the film star faced with yet another paternity suit may loathe Mendel's laws; but this is the way that the world is. Not how it ought to be, or how we would like it to be.

Even in the social sciences, such neutrality supposedly prevails. Listen to Nagel on the subject:

> There is a relatively clear distinction between factual and value judgments, and ... however difficult it may sometimes be to decide whether a given statement has a purely factual content, it is in principle possible to do so. ... [I] t is possible to distinguish between, on the one hand, contributions to theoretical understanding (whose factual validity presumably does not depend on the social ideal to which a social scientist may subscribe), and on the other hand contributions to the dissemination or realization of some social ideal (which may not be accepted by all social scientists). (Nagel, 1961, pp. 488–9)

I can only say bluntly that study of the history of science has shown me that this position is totally mistaken. It is as much a myth as in Pegasus, the Loch Ness monster, and the proposed solution of the Ruse children to the mystery of the missing cookies.

As I read into the science of the nineteenth-century, it became more and more apparent to me that everyone – creationist or evolutionist – had every one of the values or prejudices of his fellow Victorians. And, these ideas came rampantly right into the science itself. Think for the moment of some of the things the Victorians held dear. For a start, there were the claimed virtues of capitalism, certainly cherished by those who benefitted. One could hardly expect the grandson of Josiah Wedgewood, one of Britain's leading industrialists, to remain silent. And, as this passage from the *Descent of Man* well illustrates, he was not:

In all civilised countries man accumulates property and bequeaths it to his children. So that the children in the same country do not by any means start fair in the race for success. But this is far from an unmixed evil; for without the accumulation of capital the arts could not progress; and it is chiefly through their power that the civilised races have extended, and are now everywhere extending, their range, so as to take the place of the lower races. Nor does the moderate accumulation of wealth interfere with the process of selection. When a poor man becomes rich, his children enter trades or professions in which there is struggle enough, so that the able in body and mind succeed best. The presence of a body of well-instructed men, who have not to labour for their daily bread, is important to a degree which cannot be over-estimated; as all high intellectual work is carried on by them, and on such work material progress of all kinds mainly depends, not to mention other and higher advantages. (Darwin, 1871, 1, p. 169)

What about the Victorian jingoism, putting down everyone who was not born a true Briton. Remember: "Wops begin at Calais." Nor should we forget the incredible contempt for the Irish. Listen now to the popular Christian controversialist, Hugh Miller, on the subject. In a quite fantastical argument, Miller manages simultaneously to make Jesus Christ an Englishman, and to express the most unChristian sentiments about the rest of God's creation.

Now, all history and all tradition, so far as they throw light on the question at all, agree in showing that the centre in which the human species originated must have been somewhere in the temperate regions of the east, not far distant from the Caucasian group of mountains. All the old seats of civilization, – that of Nineveh, Babylon, Palestine, Egypt, and Greece, – are spread out around this centre. And it is certainly a circumstance worthy of notice, and surely not without bearing on the *physical* condition of primaeval humanity, that in this centre we find a variety of the species which naturalists of the highest standing regard as fundamentally typical of the highest races of the globe. . . .

It walks, however, the boards of our Parliament House here in a very respectable type of Caucasian man; and all agree that nowhere else in modern Europe is it to be found more true to its original contour than among the high-bred aristocracy of England, especially among the female members of the class. . . .

Let me next remark, that the further we remove from the original centre of the race, the more degraded and sunk do we find the several varieties of humanity . . . till at the extremity of the [American] continent we find, naked and shivering among their snows, the hideous, small-eyed, small-limbed, flat-headed Fuegians, perhaps the most wretched of human creatures. (Miller, 1855, pp. 250–4)

But goodness, are we not forgetting somebody?

> People who are remarkable for open projecting mouths, with prominent teeth and exposed gums; and their advancing cheek-bones and depressed noses bear barbarism on their very front. In Sligo and northern Mayo the consequences of the two centuries of degradation and hardship exhibit themselves in the whole physical condition of the people, affecting not only the features, but the frame. Five feet two inches on an average, – pot-bellied, bow-legged, abortively featured, their clothing a wisp of rags, – these spectres of a people that were once well-grown, able-bodied, and comely, stalk abroad into the daylight of civilization, the annual apparition of Irish ugliness and Irish want. (Miller, 1857)

Or, if you prefer an evolutionist on the subject, let me simply quote one Charles Darwin.

> The care-less, squalid, unaspiring Irishman multiplies like
> "rabbits: the frugal, foreseeing, self-respecting, am-
> "bitious Scot, stern in his morality, spiritual in his
> "faith, sagacious and disciplined in his intelligence,
> "passes his best years in struggle and in celibacy,
> "marries late, and leaves few behind him." (Darwin, 1871, 1, 174)

I could go on indefinitely, showing how the Victorians let their values get into their science. But, let me conclude by turning to the perennial subjects of sex and sexuality. Victorian views emerge in just amount every way, from straight statements of 'fact' to ludicrous hints, like Darwin's putting a discussion about sexuality into Latin so that children and servants could not follow (Darwin, 1871, 1, 13). To a man (*i.e.* to a *man*) there was agreement that not only are males and females different, but that science backs the view that women are simply not as bright or aggressive as men. Listen, for example to the outbursts of two of the critics of the evolutionist Robert Chambers. First we have Adam Sedgwick, Woodwardian professor of geology at Cambridge and in many respects one of the most wonderful people you could hope to meet.

> But who is the authority we thought, when we began to 'The Vestiges,' that we could tract therein the markings of a woman's foot. We now confess our error; and for having entertained it, we crave pardon of the sex. We were led to this delusion by certain charms of writing – by the popularity of the work – by its ready boundings over the fences of the tree of knowledge, and its utter neglect of the narrow and thorny entrance bywhich we may lawfully approach it; above all, by the sincerity of faith and love with which the author devotes himself to any system he has taken to his bosom. We thought that no *man* could write so much about natural science without having dipped below the surface, at least in some department of it. In thinking this, we now believe we were

mistaken. But let us not be misunderstood. Within all the becoming bounds of homage, we would do honour to the softer sex little short of adoration. In taste, and sentiment, and instinctive knowledge of what is right and good – in discrimination of human character, and what is most befitting in all the moral duties of common life – in every thing which forms, not merely the grace and ornament, but is the cementing principle and bond of all that is most exalted and delightful in society, we would place our highest trust in woman. But we know, by long experience, that the ascent up the hill of science is rugged and thorny, and ill-fitted for the drapery of a petticoat; and ways must be passed over which are toilsome to the body, and sometimes loathsome to the senses. (Sedgwick, 1845, pp. 3–4)

Then there was Sir David Brewster, Scot, optician, biographer of Newton, enemey of Whewell, and general man of science:

There is a condition of mind, the result of education and natural temperament, peculiarly open to the reception of novel and easily comprehended doctrines. Its leading feature is its impatient of that slow inductive process by which great truths are established by one mind, and through which they are demonstrated to other minds of similar character, though unequal power; and we need hardly tell our readers, that truths thus established, and thus capable of being communicated with the evidence of demonstration, are the only realities of science. The mould in which Providence has cast the female mind, does not present to us those rough phases of masculine strength which can sound depths, and grasp syllogisms, and cross-examine nature. With such a conformation, we should have lacked its soft and gentle temperament – its quick appreciation of character – and that yielding submission to a stronger nature, with which it is destined to blend. A jury of the Muses could not have administered the impartial justice of Rhadamanthus; nor could a quorum of the Graces have extricated Daedalus from his labyrinth. Hence it is that doctrines such as those of Phrenology and Mesmerism, have collected their followers chiefly from one sex; and if we have rightly gathered the rumours of the day, the most numerous and arden admirers of The Vestiges of Creation, have perused it in the boudoir and the drawing-room. It would augur ill for the rising generation, if the mothers of England were infected with the errors of Phrenology: it would augur worse were they tainted with Materialism. (Brewster, 1845, p. 503)

As always, one can rely on Herbert Spencer for the appropriate kind of comment: in his opinion, men to women were as Englishmen to savages. And then there was that well-known authority on the sexes, the author of the *Descent of Man.*

Man is more courageous, pugnacious, and energetic than woman, and has a more inventive genius. In compensation, woman has "greater tenderness and less selfishness." (Darwin, 1871, 2, 316, 326)

And then there was the matter of sexuality itself. Listen to the evolutionist Chambers, who made quite explicit his attitude to the whole matter. In the course of an argument purportedly showing that even something as revolting as evolution is not beyond the bounds of possibility, he wrote as follows:

Were we acquainted for the first time with the circumstances attending the production of an individual of our race, we might equally think them degrading, and be eager to deny them and exclude them from the admitted truths of nature. Knowing this fact familiarly and beyond contradiction, a healthy and natural mind finds no difficulty in regarding it complacently. Creative Providence has been pleased to order that it should be so, and it must therefore be submitted to. Now the ideaas to the progress of organic creation, if we become satisfied of its truth, ought to be received precisely in this spirit. (Chambers, 1844, p. 234)

With complacency, for God's sake!

Nor were other evolutionists far behind Chambers. 'Proving' that the higher the organism, the less the reproduction, Spencer (a life-long bachelor) had the following sage comments to make:

. . . undue production of sperm cells in man leads first to headaches: this is followed by stupidity; should the disorder continue, imbecility supervenes, ending occasionally in insanity. (Spencer, 1852, p. 493)

If this is not part and parcel of the general Victorian belief that unrestrained self-abuse is the quickest route to the mad-house, I do not know what is.

It is always fun to laugh at our ancestors, and there are none who lay themselves more open to ridicule than the Victorians. But, my purpose here is not to make fun of the nineteenth century. I wanted simply to show you how greatly Victorian ideology permeated Victorian science. Is there a moral here for our time? The past spurred me to ask this question, and so I turned to the present, asking if indeed my philosophical mentors had been wrong about the value-neutrality of modern science.

I found, very quickly, that they were indeed quite wrong. Today's science drips with as many hopes and aspirations, as did yesterday's science. I have an embarrassment of riches from which to make my case, so let me simply choose one example, drawn from a controversy which much interests and excites evolutionists today. I refer to the row going on at present about the nature of the fossil record and the process of macroevolution.

Most evolutionists have followed Charles Darwin in thinking that evolution is a slow, gradual process, with forms changing imperceptibly from one species to another: *Australopithecus afarensis* ⟶ *A. africanus* ⟶ *Homo habilis* ⟶ *H. erectus* ⟶ *H. sapiens*. The fact that there are many gaps in the fossil record is seen as an artifact of the random, imperfect nature of the fossilization process (Simpson, 1953; Johanson and Edey, 1981).

Recently, however, this gradual picture of the course of evolution, "phyletic gradualism," has been challenged by a number of paleontologists, who argue that the jerky nature of the fossil record is a true reflection of reality. These neo-saltationists, who call themselves 'punctuated equilibrists', argue that evolution does really proceed in jumps, from one form to another, followed by long periods of inactivity, 'stasis' (Eldredge and Gould, 1972; Gould and Eldredge, 1977; Stanley, 1979; Gould, 1980).

Here, it would not be appropriate to get bogged down in the details of the argument. I must emphasize that there is a major empirical side to the controversy, with important and fascinating studies being done of the fossil record, and of other pertinent facets of the organic world. My purpose is simply to point out that history had sensitised me to the ideology in science. Thus prepared, it was easy to find the ideology the current paleontological debate.

In particular, one can readily see that Stephen Jay Gould, a leading punctuated equilibrist, is motivated for his cause at least in part by the fact that he is a Marxist. Specifically, Gould pushes his scientific position for three Marxist-related reasons (Gould and Eldredge, 1977, Gould, 1980). First, the suddenness of change, as postulated by his vies, fits nicely with Gould's overall philosophy of the necessity of sudden, revolutionary change. By his own admission, Gould finds the gradualism of Darwinism to be a vestige of nineteenth-century British liberalism.

Second, Gould sees his position as emphasizing the historical wholeness of the organism, as postulated by Marxism. For Gould, either an organism is (say) a reptile, or it is not. We must look at things as integrated wholes; not as things which can be reduced to simple parts. Moreover, to understand an organism, we must understand its past, as well as its present.

Third, Gould likes punctuated equilibrism's immediate application to our own species, *Homo sapiens*. Either, an organism is a human, or it is not. There are no borderline cases. But, if an organism is human, then it is at one with the rest of humankind. This fits nicely with Gould's Marxist environmentalism, which sees all human differences as reflections of environmental causes, and not as innately caused by the genes. In short, Gould finds that his new biology supports, and is supported by, his old philosophy.

Please note one important point. Here, I do not condemn Gould, his theory, or his ideology. In fact, I myself favour a more orthodox Darwinian interpretation of life's history (Ruse, 1982a). But, nothing I have said here proves this point. What I do hope to have shown is that one of the most exciting hypotheses in biology today has an ideological undertone. Moreover,

I myself would say, 'Why not?' Punctuated equilibrium, even if wrong, has advanced the understanding of the fossil record, and if indeed it does prove true (*i.e.* if its predictions consistently succeed), then perhaps the time will indeed have come to start taking Gould's ideology a little more seriously. In other words, the ideology of science exists and is important whether you agree with it, or not.

With this example concluded, I have said enough on the need for the philosopher of science to study history of science. I do not pretend that every philosophical point I have made absolutely required history. History is certainly indispensible in some areas of the philosophy of science, for example, in analyses of theory change. Perhaps, history is less important in other respects. Possibly, one could arrive at the ideology of modern science without history. Feminists certainly claim that they could discern directly the sexism of Freudianism and sociobiology. Perhaps so. And then again, perhaps not. Without denying that such enterprises do have an ideology, at times I question whether the feminists have correctly identified the nature of this ideology. But, this is a matter for debate at another time. Here, I simply conclude that one philosopher of science now sees how greatly his philosophy has benefitted from his history.

PHILOSOPHY GUIDES HISTORY

What about the other direction? In what way or ways can history of science benefit from philosophy of science? Again, I see two ways, and in a sense they correspond to the benefits that philosophy receives from history.

First, most obviously and directly, one simply cannot write history of science without some sort of philosophy, whether one explicitly admits it as such or not. History is not chronology, putting 'one damn fact after another'. The totally universal, totally uninterpretive history would be Teutonically long, incredibly boring, and completely worthless. Right from the start, the historian must *select, select, select*. You have to choose the facts you are going to mention, and choose the facts you are going to ignore. And then, at the same time, you have to decide precisely what use you are going to make of your facts, and what connections you will draw. All of this, requires a philosophy.

For instance, in writing a book on the Darwinian Revolution (Ruse, 1979), I had to decide when to begin my story, and when to end it. Would I go back to the Greeks, or would I start with the Victorians? Would I end with the Victorians, or would I come forward to the present? And then, what about place? Should I confine my story to Britain, or should I go abroad also? More generally, other than science, what themes should I stress? Religion, certainly. But, what about philosophy? And, should social factors get equal treatment? A host of questions had to be answered.

I do not pretend that every answer I did give was totally arbitrary, or simply a function of my personal whim. Given the Channel, and given the Englishman's inability to speak any language but his own, I found myself rather naturally guided towards a major emphasis on the British scene. Again, it was hardly I who decided to make religion a large factor in the Darwinian Revolution. But, the personal factor did, most obviously, loom large.

And, I know that my philosophy of science influenced my decisions. Indeed, decisions had to be made which presupposed a philosophy. For instance, although I ended up by not having any explicit philosophical discussion whatsoever in the book, I was spurred into initial action by an urge to prove Kuhn wrong. I wanted to show that there are continuities in science, and that even the greatest revolutions in science do not involve total breaks. And, this urge undoubtedly guided me in my choice of figures to discuss, and in the emphases I drew. In short, the form of the book itself reflected philosophical concerns.

At this point, more critical readers may be a little worried. By my own admission, I seem to have written propaganda not history, and my philosophy and history seem to have been caught in a vicious circle: I used my philosophy to inform my history, which I then read back as justification for my philosophy! Nevertheless, although I admit there was a circle here, I deny that it was vicious. Rather, we had a self-regulating process, a phenomenon which is so common in engineering. My philosophy led me to ask historical questions; inasmuch as these came out positively, I went ahead; and when they did not, I stood back and rethought the matter.

In fact, as if in confirmation of this point, I have admitted already that history led me to modify my philosophy. I am no longer so sure of the rationality of science as I once was. Moreover, the search for continuities did not always go quite as I had planned. I found continuities, but not necessarily where I thought I would find them. I naturally assumed that one would see a nice, easy succession of evolutionists: Lamarck, Chambers, Darwin. This was not so, at all. Indeed, Darwin worked out his ideas before Chambers! The succession I sought did eventually come; but, it was between the *non*-evolutionists and Darwin! There was a line from Cuvier, Sedgwick, Whewell, Herschel, and (most particularly) Lyell, which led straight to Darwin. So, I think I still showed Kuhn wrong, but not in the way I intended to.

But, whatever the true course of events, it is indubitable that my philosophy influenced my history. Because I looked for continuities, I was led to the importance of Cuvier, and thus I gave him a more direct and extended treatment, than I would have otherwise. And, there were many other cases where similar sorts of things occurred. Moreover, what I suggest is that it was right and proper that there should have this kind of interaction. I say this, not in judgement of what I produced – what author does not think he/she could have done it better, as soon as the book appears in print? – but, simply to point out that no history can be written without judgement, and that the act of judgement in this context is philosophical.

PHILOSOPHY AS HISTORY OF SCIENCE

There is a second major way in which history benefits from philosophy. This, perhaps the most obvious of all, occurs when philosophical ideas become part of the very grist for history itself. I refer, of course, to the times when scientists of the past have used or been influenced by philosophical ideas: thus, necessarily, the historian of science must take note of these ideas. This may all seem very obvious; but, there are influential historians of science who would deny that this sort of thing ever really occurs, and so it is certainly a point worth making, even if yet again.

Perhaps the Darwinian Revolution was atypical, in that leading participants were themselves major philosophers in their own right. I refer, of course, to John F. W. Herschel (1831) and William Whewell (1840). But, be this as it may, it is undoubtedly true that philosophical factors were of major importance in the Darwinian Revolution. Let me mention three items.[3]

First, when Darwin initially came to scientific awareness, it was as a geologist that he entered the community. And, very significant this proved for Darwin's future career as a scientist, for he did not immediately get locked into minute studies of the geological record. Rather, he was forced to think seriously and deeply about the nature of science itself; because geologists were split right down the middle on their subject. Specifically, one had the followers of Charles Lyell, the 'uniformitarians', who argued that the earth is in a steady-state and that causes of a kind and intensity seen today are enough to explain the geological past. And, one had 'catastrophists', like Adam Sedgwick, who argued for major unknown cataclysisms in the past (Rudwick, 1972).

The point to be noted is that these rival geologies were not based on simple differences about fact. To the contrary, they reflected underlying differences of religion and philosophy: the deistic empiricism of John Herschel (leading to uniformitarianism), and the theistic rationalism of William Whewell (leading to catastrophism). Thus, because of philosophy, Darwin was made to ponder with care about the nature of science. Moreover, when Darwin opted for uniformitarianism – a move which led him eventually to evolutionism – he clearly did so because he was under the influence of Herschel. You may not give philosophy the significance I would give it; but, I defy you fully to understand Darwin's move to evolution, without your having a grasp of the philosophy of the 1830's.

Second, and even more crucially, you cannot understand the structure of the theory of the *Origin*, without a grasp of the philosophy. Darwin was not some bucholic bungler, who threw together all sorts of ideas, and then hoped that some of them would last. Rather, he was a skilled methodologist, who wove a careful tapestry, as he aimed to convince his audience, not only of the fact of evolution, but also of the mechanism of natural selection.

And to do this, Darwin adopted both Herschelian empiricism and Whewellian rationalism. In particular, in invoking his analogy of artificial selection, Darwin hoped to prove a Herschelian empiricist *vera causa* ('true cause'). And then, in the second half of the *Origin*, as he showed that selection could explain instinct, paleontology, biogeography, morphology, classification, embryology, and more. Darwin provided a paradigmatic example of what Whewell called a 'consilience of inductions': a rationalist *vera causa*. In short, the wiley Darwin covered his options, using both of the prominent methodologies of his day. And, unless you know some philosophy, you will miss this fact![4]

Third, the reception of Darwinism cannot be understood without philosophy. Why did 'Darwin's bulldog', T. H. Huxley of all people, never come to accept natural selection as a fully satisfying, evolutionary mechanism? The reason was revealed in a letter to his chum Charles Kingsley (Huxley Papers, 19:212). As an empiricist, Huxley felt Darwin had failed to make his case! Specifically, no one had yet created a new species through artificial selection. Thus, Huxley complained that natural selection is no true *vera causa*, and hence he withheld full assent.

Most interesting, in the light of such criticisms, Darwin turned more and more to his consilience for comfort (Ruse, 1979). Just at the time that Whewell, by now Master of Trinity, was banning the *Origin* from the shelves of the Wren Library! But, the important point to be noted is that, as in the formation of Darwin's theory, philosophical factors remained important in the *Origin's* acceptance or nonacceptance. Historians remain ignorant of these factors, or ignore them, only at their peril – and to the detriment of their history.

I conclude, therefore, that in these two major respects – the very writing of history itself and the content that the historian treats – my experience suggests that good history of science just cannot be done in ignorance of the philosophy of science. The historian needs the philosopher, no less than the philosopher needs the historian. I make no claims for the quality of what I have produced, but I would like to think that I am headed in the right direction, and I invite you to come along side with me.

In fact, you will find yourself in a growing and vigorous group, as the papers in this volume well attest. So, now, why not turn the pages, and just read on!

The answer is of course – yes. But the remedy needed is quite radical. What we must do is to replace the beautiful but useless formal castles in the air by a detailed study of primary sources in the history of science. *This* is the material to be analyzed, and *this* is the material from which philosophical problems should arise. And such problems should not at once be blown up into formalistic tumors which grow incessantly by feeding on their own juices but they should be kept in close contact with the process of science even if this means lots of uncertainty and a low level of precision. (Paul Feyerabend, 1970)

University of Guelph

NOTES

[1] Unfortunately, no publication came from the Pittsburgh conference. However, the Reno conference has since been edited, and has appeared in print (Nickles, 1980a, b).

[2] J. Beatty, *Evolution and the Semantic View of Theories*, unpublished PhD thesis, Indiana, 1978, J. R. Brown, *Models of Rationality and the History of Science*, unpublished PhD thesis, University of Western Ontario, 1981. In fairness to my good friend Ronald Giere, I must note that he was one of the supervisors of Beatty's thesis, and, for all his published statements, actively encouraged work on the historical aspects of evolutionary biology. Expectedly, Robert Butts supervised Brown's thesis.

[3] Any work I myself did in this particular direction was merely footnoting pioneering achievements by Michael Ghiselin (1969) and David Hull (1973).

[4] This is my version of accounts, which I expand on in Ruse (1979). At the Montreal conference, John Hodge gave a different analysis, challenging my interpretation. But, we both agree that philosophical elements crucially influenced Darwin's position in the *Origin*. As Robert Butts notes, we hope that Hodge's analysis will shortly be forthcoming as an independent monograph.

BIBLIOGRAPHY

Braithwaite, R. B.: 1953, *Scientific Explanation*, Cambridge Univeristy Press, Cambridge.

Brewster, D.: 1844, *Vestiges, etc., North British Review* 3, 470–515.

Bunge, M.: 1968, 'Analogy in quantum theory: From insight to nonsense', *Brit. J. Phil. Sci.* 18, 265–86.

Chambers, R.: 1844, *Vestiges of the Natural History of Creation*, Churchill, London.

Darwin, C.: 1859, *On the Origin of Species*, Murray, London.

Darwin, C.: 1871, *The Descent of Man*, Murray, London.

de Beer, G.: 1963, *Charles Darwin: Evolution by Natural Selection*, Nelson, London.

Eiseley, L.. 1958, *Darwin's Century*, Doubleday, New York.

Eldredge, N. and Gould, S. J.: 1972, 'Punctuated equilibria: an alternative to phyletic gradualism', in T. J. M. Schopf (ed.), *Models in Paleobiology*, Freeman, Cooper, San Francisco.

Feyerabend, P.: 1970, 'Against method: Outline of an anarchistic theory of knowledge', in M. Radner and S. Winokur (eds.), *Minnesota Studies in the Philosophy of Science* 4, 17–130, University of Minnesota Press, Minneapolis.

Ghiselin, M.: 1969, *The Triumph of the Darwinian Method*, University of California Press, Berkeley.

Giere, R.: 1973, 'History and philosophy of science: an intimate connection or a mar riage of convenience'? *Brit. J. Phil. Sci.* 24, 282–9.

Gould, S. J.: 1980, 'Is a new and general theory of evolution emerging?' *Paleobiology* 6, 119–30.

Gould, S. J. and Eldredge, N.: 1977, 'Punctuated equilibria: the tempo and mode of evolution reconsidered, *Paleobiology*', 3, 115–51.

Hempel, C.: 1966, *The Philosophy of Natural Science*, Prentice-Hall, Englewood-Cliffs.

Herschel, J. F. W.: 1831, *Preliminary Discourse on the Study of Natural Philosophy*, Longman, Rees, Orme, Brown, and Green, London.

Hull, D. L.: 1973, *Darwin and His Critics*, Harvard University Press, Cambridge, Mass.

Johanson, D. and Edey, M.: 1981, *Lucy: The Beginnings of Humankind*, Simon and Schuster, New York.

Kordig, C.: 1971, *The Justification of Scientific Change*, D. Reidel, Dordrecht, Holland.

Korner, S.: 1960, 'On philosophical arguments in physics, in E. H. Madden (ed.), *The Structure of Scientific Thought*, Boston.

Kuhn, T.: 1962, *The Structure of Scientific Revolutions*, University of Chicago Press, Chicago.

Miller, H.: 1857, *The Testimony of the Rocks*, Constable, Edinburgh.

Nagel, E.: 1961, *The Structure of Science*, Routledge and Kegan Paul, London.

Nickles, T.: 1980a, *Scientific Discovery: Logic and Rationality*, D. Reidel, Dordrecht, Holland.

Nickles, T.: 1980b, *Scientific Discovery: Case Studies*, D. Reidel, Dordrecht, Holland.

Rudwick, M. J. S.: 1972, *The Meaning of Fossils*, Macdonald, London.

Ruse, M.: 1979, *The Darwinian Revolution: Science Red in Tooth and Claw*, University of Chicago Press, Chicago.

Ruse, M.: 1980, 'Ought philosophers consider scientific discovery? A Darwinian case-study', in T. Nickles (ed.), *Scientific Discovery*, D. Reidel, Dordrecht, Holland.

Ruse, M.: 1981, *Is Science Sexist? And Other Problems in the Biomedical Sciences*, D. Reidel, Dordrecht, Holland.

Ruse, M.: 1982a, *Darwinism Defended: A Guide to the Evolution Controversies*.

Ruse, M.: 1982b, 'The revolution in geology', in P. Asquith (ed.), *PSA 1978*, 2.

Scheffler, I.: 1963, *The Anatomy of Inquiry*, Knopf, New York.

Scheffler, I.: 1967, *Science and Subjectivity*, Bobbs-Merrill, Indianapolis.

Sedgwick, A.: 1845, *Vestiges, Edin. Review* 82, 1–85.

Simpson, G. G.: 1953, *The Major Features of Evolution*, Columbia University Press, New York.

Spencer, H.: 1852, 'A theory of population, deduced from the general law of animal fertility', *Westminster Rev.* 1, 468–501.

Stanley, S. M.: 1979, *Macroevolution: Pattern and Process*, W. H. Freeman, San Francisco.

Whewell, W.: 1840, *Philosophy of the Inductive Sciences*, Parker, London.

PART I

ROBERT JOLY

HIPPOCRATES AND THE SCHOOL OF COS

BETWEEN MYTH AND SKEPTICISM

The myth is well known: it is dispersed throughout the apocryphal texts of the Hippocratic *Collection*, the *Letters, Decrees* and *Speeches* (Littré, 1861, IX, pp. 308–428; Hercher, 1871, pp. 289–318; and Putzger, 1914): the king's invitation, the patriotism of Hippocrates, the story of his relation to Democritus,[1] the role he played at the time of the plague in Athens. The Hippocratic fervour that this heroic and scientific gesture aroused has not entirely dissipated since contemporary works for the public at large – perhaps exclusively francophone – retain significant aspects of it or weave new fantasies alongside these themes (see Joly, 1966, p. ff.).

However, since specialists have unanimously recognized it as such for a very long time, the myth, in spite of these aberrant resurgences, has lost all its virulence.

But there is a mythic aspect distilled from it, more circumspect and more insidious, with no link to the apocryphal writings, which persists in the heart of specialized research. It is the valorization of Hippocratic medicine, of the *Collection* in general, and of the School of Cos in particular. In a frequently spontaneous and unreflective way, outside the bounds of a specific epistemological framework in any case, Hippocratic medicine is conceived of as very close to contemporary medicine, sometimes as superior to it in certain aspects, to the point where a return to Hippocrates, much desired and celebrated by several colloquiums on Hippocratic medicine seems to some people to be in order (see Bourgey, 1953, p. 275; and Joly, 1966, p. 10 passim).

This belated modernisation results from the conjunction of several factors. The emotional attachment to the Father of Medicine, the vague nostalgia for all 'Back to . . . " movements has been inadequately opposed by critical attacks for prosaic reasons: the Greek philologist concerned with the *Collection* does not know enough about modern science and thinks therefore that he has rediscovered it in the Greek texts he is studying for the greater glory of Hellenism for which he is always something of a missionary. For his part, the physician who knows, or ought to know what science is, is not often a trained historian, so that when reading Hippocrates he projects his own

Michael Ruse (ed.), Nature Animated, 29–47.

knowledge into what he reads.[2] For lack of critical vigilance the Hellenist and the physician, too easily victims of the evangelical message, often believe they have found what they were looking for, what they more or less unconsciously want to find.

In 1966 I devoted a book to this other Hippocratic myth, I have returned to the subject elsewhere and quite recently (see e.g. Joly 1966, 1972a, 1980) in order to reply to published objections. I have nothing to add at the moment but it will be admitted perhaps that my position was not felt to be unduly credulous.

In any case present circumstances have turned my attention to the opposing standpoint which seems to me to be one of excessive skepticism.

In our field of studies, the most radically critical attitude in the preceding generation was that of Ludwig Edelstein (1931). He was a great scientist, a scholar with often novel views who brought a great deal to the history of ancient medicine. But as far as the *Collection* was concerned and its links with Hippocrates, he was an extreme nihilist. Wilamowitz had written that Hippocrates was a name without a book: Edelstein (1935, col. 1328) outdid him: "Hippocrates is a name lacking any accessible historical reality."

This extreme skepticism has been the object of many, but rather diverse clarifications. Personally, I consider K. Deichgraeber's study (1934, 2nd ed. 1971) as admirable a synthesis of the School of Cos as a sound critical mind could attempt at the time. Certainly many questions remained unresolved[3] and Deichgraeber himself (1934, p. 170 passim) indicated research that, following his lead, it would be important to carry through to a successful conclusion.

Since then, and in spite of a later article by Edelstein in the *Realencyclopedie* (1935) very little more was heard of the radical tendency. Hippocratic research devoted itself to the minutiae of strictly limited enquiries and often raised questions. The very broad syntheses did not follow in Edelstein's footsteps but inclined rather toward an excessive valorization. (See e.g. Bourgey, 1953; P. Lain Entralgo, 1970).

A point of view very close to that of Edelstein, on the other hand, has just appeared in Wesley D. Smith's book, *The Hippocratic Tradition* (1979). This is not a study of the *Collection* itself but of the history of the Hippocratic tradition from the 4th century B.C. to the present.

The focal point of the book is the chapter on Galen. Smith knows Galen well[4] but he denounces his evidence concerning Hippocrates too categorically as rhetorical and self-interested assertions upon which historians have mistakenly relied.

For my part I am ready to recognize that too much confidence has sometimes been placed in Galen. I would also agree very broadly with Smith in his first chapter which outlines the history of the Hippocratic tradition since the Renaissance and shows the role of Littré and his influence right up to the present. But on the studies of the last thirty years Smith devotes a scant few pages (mainly p. 43) and shows a marked disdain for the 'current work' being done.

The essential point however is that between the interest in Hippocrates of the Empirics, who were the first to found the Hippocratic tradition, and Hippocrates himself, Smith discovers nothing but a complete desert. No one before the first Empirics attached the least importance to Hippocrates or to works considered to be Hippocratic, no one claimed or rejected him as a real authority in medical matters (see Smith, 1979, pp. 178, 199, 204, 208).

It is here that it seems to me there is an excessive degree of skepticism. Even the fact that Smith has foreseen the reproach (Smith, 1979, p. 219) does not dissuade me: one must look at the facts themselves, and draw from them, with the strictest severity, the only conclusions they allow. In my opinion the harvest is, after all, much richer than Smith's fruitless gleanings. And this is what I would like to show in the following pages, limiting myself, deliberately, to the most ancient testimonies, some of which do not even appear in Smith's book, probably because they have been judged too meagre; others are touched upon but abandoned before having delivered up all the information contained in them.

In the *Protagoras* (311 b–c), a youthful dialogue, Plato places "Hippocrates of Cos, one of the Asclepiads," as physician on the same footing as Polycletus and Phidias as artists. We also learn in this valuable passage that he has paying students.

It must be concluded from this text that Hippocrates is indeed a celebrity and even *the* celebrity in medical circles of the day,[5] and if this is indeed his reputation in Athens it is because he travels a great deal, and probably also because he writes and publishes.

L. Edelstein (1935, col. 1325) cleverly opposes to these conclusions a passage from *Phaedrus* by the same Plato where two very little known physicians are cited in the same breath with Sophocles and Euripides and where one can certainly not conclude that Erixymachus and Akoumenos would be celebrities comparable to the two great dramatists.

But it must be recognized that these physicians are only mentioned by virtue of their close relationship to Phaedrus, whereas in the *Protagoras* nothing links Hippocrates to the young man who comes in search of Socrates, beyond the coincidence of their common name. In addition the brevity of the passage firmly places Hippocrates in close connection with Polyclitus and Phidias, which is not the case in the *Phaedrus*.

In his *Politics* Aristotle wrote for the generation to come: "About Hippocrates one could say that the physician and not the man is greater than he who exceeds him in size (VII, 1326 a 15–16)." Edelstein wants to conclude from this text only that Hippocrates was a small man (1935, col. 1297) something that is confirmed in the Brussels manuscript *Life.*[6] But it is also obvious from this sentence that the name Hippocrates, here used alone, is sufficient to make clear of whom one is is speaking, and furthermore that he is great as a physician: he is still, in Aristotle's time, the greatest celebrity of the medical world. L. Bourgeys' analysis is faultless and it would be pointless to wish to disregard it. "Such a person has already in some fashion entered history and the ranks of the famous. It is worth noting also the attribute of greatness given to Hippocrates without any discussion, as something self evident. Aristotle has a practical mind, little given to excessive praise and as the son of a physician into the bargain he must know whereof he speaks." (Bourgey, 1953, p. 83)

That Hippocrates was known as the author of published works seems to me implicitly confirmed by Plato's *Phaedrus*. Whatever interpretation one accepts of the famous passage (270 b–c) it questions a Hippocratic doctrine that can only come from a written work or works. Can one imagine Plato taking notes at a lecture given in Athens by Hippocrates? The conclusion is that this kind of *epideixis* was intended for publication.

Ctesias' evidence points in the same direction and in addition offers us a strong probability in connection with one work in particular.

Galen (XVIII A 731 Kühn) informs us that physicians have criticized Hippocrates for his reduction of the luxation of the thigh; "The first one was Ctesias, his relative, who was himself a member of the Asclepiad family, and after Ctesias others as well." To eliminate this piece of evidence it has to be judged implausible: "The source of the strange statement that Ctesias (fifth century B.C.) criticized Hippocrates on the subject and that, as an Asclepiad, he was Hippocrates' relative, is difficult to image (Smith, 1979, p. 128)." However, the most surprising detail in this sentence is confirmed by an inscription at Delphi which relates precisely a decision taken by the Koinon of the Asclepiads of Cos and Cnidus, the homeland of Ctesias (see J. Bousquet, 1956, p. 579). In addition, one has to impute Galen's statement to his desire to find references to Hippocrates everywhere. "But it is like Galen to populate history with quarrels against Hippocrates." (Smith, 1979, 1.1. and p. 180; also cf. Lloyd, 1975, p. 176).

Since it is difficult all the same to accept that Galen fabricates one hundred per cent of the time, it is thought that Ctesias is alluding or could be alluding to an earlier work and that it is Galen who, identifying the designated procedure in a passage of *Joints*, attributes to Ctesias a criticism of Hippocrates.

We certainly know that there existed in the fifth century a vast medical literature, but our surest proof of it, the author of the *Regimen*, talks of dietetic works and we know from other sources as well that preoccupations of this kind are very ancient in Greece (see Joly, 1967, p. xi passim). But on a subject as limited and technical as *Fractures-Joints*, the situation may be very different and the probability is, I think, quite the opposite. And then to accept that another work had shown the same reduction considered however by Ctesias to be useless does not increase the probability.

It seems to me that one cannot 'ignore' (Smith, 1979, p. 179) Galen's evidence concerning Ctesias except by giving proof of bias. I do not claim that the passage furnished proof but it is difficult to accept that it does not provide a strong probability.

This seems all the more certain since, for the succeeding generation, we have unimpeachable proof that the treatise in question was read.

Galen, commenting on an old fashioned term *τύρσις* quotes what he himself calls a paraphrase from a Hippocratic passage by Diocles of Carystus, a paraphrase which he gives us word for word and where two ancient terms from the text have been modernized (XVIII A 519 K. and see L. Bourgey, 1953, p. 100, n. 3). It is worth noting, incidentally, that this time, and in spite of the desire Smith ascribes to Galen, the latter does not explicitly attribute to Diocles a reference to Hippocrates himself.

It is clear that in the middle of the 4th century [7] *Joints* was read, in spite of the time lapse and the antiquity of the language. Why, since then, would Galen invent when he shows us Ctesias, a contemporary of Hippocrates, reading and criticizing the work? The passage from Diocles constitutes Wellman's fragment 187 but, as far as I can see, Smith does not make use of it.

If one turns to the *Epidemics I and III* the probability of their attribution to Hippocrates himself is even stronger than for *Fractures-Joints*.

This work is dated very positively about 410. It takes us to Thrace and the adjacent islands, and both the history and the epigraphy, of Thasos in particular, provide accurate across checks (see J. E. Dugand, 1977, pp. 233–245). On the other hand, Epidemics II, IV, and VI closely related to I and III date from the beginning of the 4th century and take us to Thessaly. Larissa where, according to the Lives (Soranus, Chapter 11), Hippocrates tomb is located is not named[8] but rather Pharsalus and Kranon.

It is difficult to escape these parallels. Even Edelstein admits that Hippocrates died in Thessaly (1935, col. 1297).

To go any deeper into skepticism one would have to deny any truth to the *Lives* of Hippocrates.

Certainly the *Lives* do include myth but that need not mean at all that on factual questions about immediate relationships of date or of place they are devoid of value. A person as well known as Hippocrates during his own lifetime (and that much is certain) could not have been completely unknown to succeeding generations, in Cos in particular, to the point that elements of his biography, even the most neutral ones, had to be invented later on.

Then again these *Lives* clearly belong to a school of serious and scholarly biography. The *Life* of Soranus proceeds methodically and quotes numerous ancient sources and this precludes the apocryphal myth as the sole source of the lives.

Smith wants to believe that the biographical indications in the *Letters* are repeated in the *Epidemics* themselves, so that the crosscheck of the facts is pure illusion (Smith, 1979, p. 219). He is satisfied moreover, on this point as on others, with a very general suggestion whereas what is required here is a detailed examination.

If one looks for the facts common to the *Epidemics* and to the apocryphal texts, one finds practically nothing beyond the mention of Thessaly (Littré, 1861, IX, pp. 402–404). If the fabricator had wanted to exploit the *Epidemics* he could have gathered an ample harvest of geographical, onomastic, and other facts. Clearly he did not do so.

It is the same for Hippocrates near relatives. Because a genealogy going back to the gods is legendary, this does not make it legendary as well when it goes from grandfather, to father, to the person's children.

The list of the bishops of Rome may and must be suspect as far as Peter himself and his immediate successors are concerned, where the order varies according to traditions. This is not a reason to reject the list as it concerns the 2nd century and more particularly the latter half of the 2nd century. The closer we get to the period when for doctrinal reasons the list was elaborated, the closer we are to history proper.

It is unthinkable that a celebrity like Hippocrates left no memories, even at Cos, of his ancestors and his immediate descendants.

In any case the name of his son Thessalus, appears in an inscription at

Delphi (Bousquet, 1956, p. 586) while it is missing from the ancient Hippocratic *Collection.*[9] But it is his son-in-law Polybus who is of primary interest in the *Collection*: I will return to this shortly.

The apocryphal myth itself comes out of Cos and has made use of obvious Coan facts. Edelstein conceded this too (Edelstein, 1935, col. 1301). Probability alone firmly indicates Cos as the homeland of the Hippocratic myth. A very sure cross check is the appearance of a rare and technical formula at the same time in Thessalus'[10] speech, in the inscription at Delphi which concerns the Koinon of the Asclepiades, and Coan inscriptions. (See Bousquet, 1956, p. 587 and n. 1).

This is why it is not naive to retain in the mythic biography some of the more neutral and positive elements preserved in it. It is not the person who calmly sorts the material who is being ingenuous but the one who carelessly insists on all or nothing.

A few facts seem to me to resist quite well hypercritical attacks and they are decisive for the history of classical medicine. If it has to be conceded to Lloyd (1975) that there is no absolute certainty concerning the attribution of a work in the *Collection* to Hippocrates himself, one must however add that as far as *Fractures-Joints* is concerned the attribution is highly probable and even more so as far as *Epidemics I and III* is concerned.

The myth, the valorization and the projections of Galen and others are grafted onto a solid historical core that remains accessible to us in spite of the scarcity of information. To deny this core against all sound historical criticism, to wish, in spite of the evidence, to reduce Hippocrates to the modest ranks of an ordinary physician is to run the risk of no longer understanding why the legend seized on him and not another. The legend is far better explained if we begin with an exceptional person; we only borrow from the rich. Excessive skepticism makes the further mistake of transforming the legend into a well-nigh incomprehensible mystery.

* * *

And now the great surprise.

Smith, who accepts none of the preceding, holds nevertheless to one quasi-certainty about Hippocrates: that he must be the author of the *Regimen*. He devotes about twenty pages to a demonstration of this extremely original thesis (Smith, 1979, pp. 44–60). In this radically critical attempt these pages constitute a genuine anomaly.[11] Among so many negations we must examine whether this affirmation is worth retaining.[12]

Two preliminary remarks.

It must be clearly seen that if Hippocrates was the author of the *Regimen* he could not also be the author of *Epidemics I* and *III* nor of *Fractures-Joints*. There would be no chance at all that he was very close to these works and others held to belong to the School of Cos (see Joly, 1961a). The *Regimen* is, it is true, an eclectic work that finds its inspiration in many presocratics, but it is also a work where rigid medical doctrine is derived from a dualist cosmology and anthropology in which water and fire are two fundamental elements. All this could not be more directly opposed to the School of Cos to the point where it would be useless to imagine an evolution from one of the Coan works to the *Regimen* or vice versa. This would be assuming the right to adopt the most improbable hypothesis of all.

In the second place, Smith's thesis supposes a certain interpretation of the famous passage of the *Phaedrus* on Hippocrates. I cannot repeat my interpretation here (see Joly, 1961a) but I must state that I stand by it, in spite of the recent article by Jouanna (1977a) which I shall examine at another time.[13]

In any case, Smith's thesis touches only slightly on the general interpretation of the text of the *Phaedrus* and it will be enough if I note here my disagreement.

In *Phaedrus*, 270 a 1 *μετεωρολογία* is translated in a purely metaphysical sense 'lofty thoughts' (Smith, 1979, p. 45): I think that this meaning has scarcely any justification and that the primary meaning of the term is required here (see Joly, 1961a, pp. 81, 82).

As for *τῆς τοῦ ὅλου φύσεως* (*Phaedrus* 270 c 2) Smith does not resolve the question of knowing in what sense *τὸ ὅλον* is used: the Whole (= the Universe), the whole (of the object in question); he thinks that Plato intentionally suggests the two meanings at the same time, he even adds a third possible meaning: 'the nature of all body' and thinks that Plato is purposely ambiguous.

The context in which a word is placed in fact demands the exclusion of different possible meanings in favour of only one, and when the context is difficult it is up to the philologist to study it as closely as possible. Sound linguistics quite normally requires the resolution of the question, not accepting deliberate ambiguity except for very cogent reasons.[14]

Smith's essential argument can be put in a few words.

In the context of the passage from the *Phaedrus*, it is a question of the dialectical method of division *διαίρεσις*, and collection *συναγωγή*. The beginning of chapter two of the *Regimen* reads:

I contend that whoever is going to write properly about regimen for men must first know and distinguish (γνῶναι καὶ διαγνῶναι) the nature of man in general: He must know from what things man is composed from the beginning, and must distinguish the parts by which he is controlled. For if one does not know the original composition he cannot know what results from those things. And if he does not know[15] what is to control the body, he cannot know how to administer what will benefit a man.[16]

In the distinction between γνῶναι and διαγνῶναι one is to believe that Plato has discovered, at least in outline, his dialectical method of division and collection.

First of all I think that in spite of the context Plato does not attribute his dialectical method in any way to Hippocrates.[17]

Plato would have to have been extremely naive and inattentive, it seems to me, to have found or rediscovered his method in the text of the *Regimen*.

Division and collection concern, necessarily, the same object, a collection of concept-ideas that must at the same time be divided and successfully reassembled. In the *Regimen* the questions introduced successively by γνῶναι and διαγνῶναι are different.

And then γνῶναι is much too commonplace a verb to designate by any stretch of meaning the platonic συναγωγή. Smith wants to translate it as 'know together' (Smith, 1979, p. 48) but this is obviously to reinforce his case.

Furthermore, in continuing the reading of chapter two one cannot help noticing that the difference between γνῶναι and διαγνῶναι is very slight and that the variation seems above all a matter of style. All of the passage cited is summed up immediately afterwards by γινώσκειν. (Joly, 1967, p. 2, 1.16). The author then uses ἐπίστασθαι, then διαγινώσκειν, then γινώσκειν again, but the whole is repeated in διαγνόντι (see Joly, 1967, p. 2, 1.19; p. 3, 1.4–5; 1.13 and 1.17–18).

Finally in the entire work there is a frequently repeated formula, which introduces different chapters οὕτω χρὴ διαγινώσκειν (see Joly, 1967), pp. 28, 1.6; 36,2; 37,16; 45,17). But one finds almost as often οὕτω or ὧδε χρὴ γινώσκειν (Joly, 1967) pp. 38,4; 40,14; 61,16). There is not the least substantial relationship, in fact, between these passages in the *Phaedrus* and the *Regimen* and hence nothing that warrants the conclusion that when talking about Hippocrates Plato was thinking of this text.[18]

The second argument consists of the assertion that the account of the Hippocratic doctrine in the *Anonymus Londinensis* is drawn from the *Regimen*. The decisive factor is said to be that Hippocrates attributes the cause of diseases to the φῦσαι, a doctrine that is only found in the *Regimen* in Chapter 74 (Smith, 1979, p. 54).

As far as the Hippocrates of Aristotle-Menon is concerned, it must be said that the doctrine of φῦσαι, causes of diseases in general, forms part of a whole that is not found in the *Regimen*: the φῦσαι arise from περισσώματα, themselves brought on by three processes readily distinguished according to the volume or the nature of the food; and diseases arise in their diverse forms from the diversity of the φῦσαι.

As for the *Regimen*, the notion of φῦσαι plays only an extremely precise role in connection with one of the fifteen cases of prodiagnosis, that is to say in states of disequilibrium between food and exercise, conditions that can be corrected before the appearance of diseases. It is therefore inappropriate to say in this case that φῦσαι provokes disease.

The essential doctrines of the *Regimen* are obviously not found here. It would be all the more unusual to believe that the *Regimen* inspired the Hippocratic method since its doctrine is found much more clearly elsewhere in the same work: in the section devoted to Herodicus of Selymbria, who must be one of the teachers thinking of the author of the *Regimen* (Joly, 1961b, p. 203).

Clearly anticipating objections, Smith states that if the author does not mention the exercises so dear to the *Regimen* it is because "Menon mentioned that factor earlier, in his report of Herodicus and did not need to repeat it." (Smith, 1979, p. 54) This would be a strange way of conducting the profession of doxography; not to mention the exercises again because one has already mentioned them in connection with someone else. The author amply demonstrates that he is not afraid of repetition. If it concerns Herodicus of Cnidos, who is reported earlier, it is true, one only finds there the verb ἀκεινήσαντες which might refer, very vaguely to the exercises. It is in connection with Herodicus of Selymbria, who is reported later, that the question of πόνοι explicitly arises.

Other arguments are even weaker.

When Plato writes: Hippocrates and true reason (*Phaedros*, 270 c end) Smith would like to believe that this λόγος alludes (discreetly) to Heraclitus. Plato would still be thinking of the Hereclitianism of the *Regimen* (Smith, 1979, p. 48). But λόγος does not belong exclusively to Heraclitus and the ἀληθὴς λόγος of this passage is the scientific method of Plato himself. On the other hand the influence of Heraclitus on the *Regimen* has been highly exaggerated. I found for my part that it is limited more or less to one passage in Chapter five and does not play the least doctrinal role in the one hundred and ten pages of the work (See Joly, 1961b, p. 89).

Smith's thesis, in my opinion, lacks any real foundation. What can be appealed to in favour of *Epidemics I* and *III* and even *Fractures-Joints* is considerably more persuasive.

* * *

Leaving aside the person of Hippocrates we now turn to the School of Cos.

If one seriously believes that Hippocrates is the author of the *Regimen*, the problem of the School of Cos becomes an impenetrable mystery and not worth further discussion.[19] The *Regimen* is such a specialized work in the *Collection* that it is impossible to see what other works would be attributable to the medical milieu from which it emerged. On the other hand a series of other works cohere very naturally around *Epidemics* and *Fractures-Joints*: the *Prognostic, Humors*, the *Surgery, Mochlikon, Airs, Waters, Places*, the *Sacred Disease*, the *Nature of Man*.

This is not a groundless affirmation: the close links that unite these works, with some subtle differences, have been demonstrated by the analyses of K. Deichgraeger (1934, pp. 17 passim; 75 passim). One cannot deny the relationship of these works without systematically disputing Deichgraeber's work. But no one has undertaken a criticism of this kind[20] which I believe would be bound to fail.

One of the works merits our particular attention. The *Nature of Man*. I am not as convinced as Jouanna that it forms an obvious literary unity but the unity of authorship seems to me undeniable (see Jouanna, 1969, pp. 150–157).

The *Nature of Man* is attributed to Polybus by Aristotle as far as Chapter 11 is concerned, in the *History of Animals* (HA, III, 3, 512a–513b) and as far as the first chapters are concerned in the *Anonymus Londinensis.*[21]

In both cases Polybus is cited only by name: proof of certain notoriety. In the *History of Animals*, his system of the blood vessels comes at the end. Jouanna is right to see in this proof that for Aristotle Polybus is still current (Jouanna, 1975, p. 55). In the work of Aristotle-Menon, most physicians are named with an indication of their origins: Hippocrates, Plato and Polybus are almost the only ones called simply by their name.[22]

We know on the other hand, thanks to the apocryphal texts and to Galen that Polybus was a student and son-in-law[23] to Hippocrates. I have already said that there is no reason to doubt these specific facts in view of the unquestionable fame of Polybus and the Coan origin of later information.

There is no doubt at all that *Nature of Man* is a Coan work and its links with other works know to be Coan from other sources only serve to reinforce the whole.

Smith does what he can to escape these arguments. After having relegated Polybus to later consideration he writes:

Polybus is cited in the Doxography of Menon, Chapter 19, as author of a theory of nature of man and disease that seems very near to that of *Nature of Man*, chapter 8. The papyrus of *Anonymus Londinensis* is especially tattered at chapter 19. Only the first part (about half) of each line of text is legible. Diels' reconstruction of the missing text was based on *Nature of Man* and his results have since been used to prove that *Nature of Man* is the source of Menon's report of Polybus. These circular arguments should probably arouse our suspicion but, if we repress it . . . (Smith, 1979, p. 220)

When one is sure of a good half of a line of a papyrus, reconstruction of the text attains a high degree of certainty. But over and above this, to cast suspicion is far from adequate as good methodology: It would be necessary to show that another reconstruction is more plausible and also, under the circumstances, that it detaches the Aristotle-Menon text from any link with the *Nature of Man*. I maintain that it would be impossible to make both of these points and it is up to Smith to accept the challenge.

Apart from these difficulties, he has to convince himself that the uniting of Polybus, author of the *Nature of Man*, and Polybus, student and son-in-law of Hippocrates was probably the creation of Dioscurides and Capiton, deceived perhaps by an unfortunate similarity of name (Smith, 1979, pp. 220, 221). This is to prefer skeptical conjecture to some solid facts that can be seriously corroborated.

It is appropriate here to consider the more subtly shaded views of Lloyd who stressed the doctrinal divergences separating some works from certain others that have been accepted here as of Coan origin. (Lloyd, 1975, p. 184 passim)

These divergences, to Lloyd's way of thinking, show that these works are difficult to attribute to a sole author; they do not aim, it seems, at dissipating the notion of the School of Cos, a subject which, unless I am mistaken, Lloyd does not go into.[24]

In the three cases analysed by Lloyd, there is the notion of prognosis. Between *Epidemics I* and *III* and *Prognostic*, the differences are of little importance in view of the obvious relationship, but *Airs, Waters, Places* brings in only external factors and seems even, in chapter II, to exclude any other factor.

For my part, I do not maintain that *Airs, Water, Places* is as clearly by the same author as the other two works. It is enough for me that the three are from the same school and probably from the same generation.

It must be added however, that *Airs, Waters, Places* is specifically devoted to external factors and that once launched on this theme, the author may have forced his thinking a little. The fact that *Airs, Waters, Places* is preoccupied with etiology does not necessarily separate it from *Epidemics I* and *III* either. It is obvious that the author of the latter deliberately sets aside this aspect of things from his specific theme.

A second case concerns the dietetic concepts of the *Regimen in Acute Diseases* and *Ancient Medicine*: it never occurred to me that these works could be by the same author. Furthermore, and above all, *Ancient Medicine* is a difficult work to class, in spite of Littré's belief that he could reconstruct the School of Cos and the works of Hippocrates from it. K. Deichgraeber's determined silence about this work tell us a great deal.

Only the first example developed by Lloyd could worry us any further: the question of the divergences between Epidemics I and III on the one hand and the *Prognostic* on the other concerning the theory of critical days.

Placed beside the relationships between these works, recognized by Lloyd, this difference seems to me to carry little weight.

These numerological theories are no more scientific in the contemporary sense of the term than the mentality of the physicians, which means that the idea of contradiction would not have been, at the time, as rigorous and as serious as it is today. We are confronting "variations on the same theme" that should not be overly dramatized. It should not be forgotten, and Lloyd reminds us, that the *Prognostic* proposes two series of critical days, one of even days 4th, 6th, 8th, 10th, 14th, 20th, 24th, 30th, 40th, 60th, 80th, 120th and another, scarcely any shorter of uneven days. It would be wise to remember what G. Bachelard wrote about mathematical traps.

> Excesses of precision in the realm of quantity correspond exactly to excesses of the picturesque in the realm of quality. Numerical precision is often a battle of figures, just as the picturesque to borrow from Baudelaire, is a battle of details. (Bachelard, 1947, pp. 212, 213)

Along the same lines, Lloyd remarks that within *Epidemics I* itself there is some disparity on this same point between chapters 9 and 12. But he does not explicitly draw the conclusion, as the whole of his exposition would lead him to do, that *Epidemics I* is the result of the collaboration of at least two authors. In my opinion this would be excessive, the preceding considerations adequately explain the situation.[25]

If these remarks are considered valueless, it must still be said that a difference at this level can probably also be explained by a certain lapse of time separating the works of the same author, or even notes meant for the same work. We are in no position to decide with complete certainty that two works cannot be by a same author: doctrine can evolve – especially at this prescientific level – language can evolve.[26] This does not mean however that anything is possible: it will doubtless be admitted that what separates *Prognostic* from the *Regimen* has nothing in common with what separates it from *Epidemics I* and *III*.

If it is possible in this way to put one's finger on a large core of Coan works, it is also possible to arrive at the same result for a core of Cnidian works. In this type of problem that is the least of the matter: the attribution of still other works in the *Collection* being sometimes difficult and disputed.

Galen quotes some lines from Euryphon, the great Cnidian teacher of the mid-fifth century. Now these few lines can be found almost word for word in *Diseases II*. Galen is well aware of it for he quotes the passage from *Diseases II* (see Jouanna, 1974, pp. 18, 19; and H. Grensemann, 1975, fr. 15) immediately after Euryphon, but attributes the work to Hippocrates. Obviously we do not have to fall into this latter error, but we do have to consider *Diseases* as Cnidian. And this brings in *Internal Affections* and quite a few other works that it is not necessary to list here, as also belonging to Cnidian writings.

It is often forgotten that we have on this point a very ancient and very important corroboration.

The Regimen in Acute Diseases in criticizing the Cnidians in its prologue gives in fact a genuine quotation: it reproaches them for recommending almost exclusively "to give to drink, in the proper season, whey and milk" (*καὶ ὀρὸν καὶ γάλα τὴν ὥρην πιπίσκειν* Joly, 1972b, p. 36). These very words constitute a regular formula found again in precisely those works declared to be Cnidian in terms of the quotation from Euryphon,[27] and nowhere else.

It is therefore historically certain that there existed at Cos and at Cnidos a centre where medicine was taught under well known masters and where medical works were published. No one really doubts these facts, the difficult question lies elsewhere.

The existence of two schools has been thought of as a rivalry, based on profound doctrinal disagreements. The whole of modern criticism has always tended to radically separate their doctrines into concepts of medicine belonging to Cos or to Cnidos,[28] often in the desire to elevate Hippocrates and his school in relation to an archaic, mediocre, bumbling Cnidian school.

On this Smith's criticism seems to me very sound. He had published an article (Smith, 1973) before the book referred to here, in which he shows that Galen is not at all conscious of this opposition of the schools, that he never thinks of them as separated by a wide gap either in theory or in medical practice.

I accept Smith's analysis but it seems to me one question remains unanswered. In his article, but even more in his book, Smith teaches us, above all things, to mistrust Galen's evidence. We should not be satisfied therefore with his opinion but try to prove it in the texts themselves of the *Collection*. Since we have Coan and Cnidian works (but it is possible Smith goes so far as to doubt this) let us compare them directly.

I personally have done at least a part of this work and for more than fifteen years I have been writing and repeating something that seems obvious to me. that the differences which place the two schools in opposition are highly exaggerated, that what separates them arises rather out of vocabulary or turns out each time to be a tempest in a teapot.[29] My position was quite isolated at the time and that is changing somewhat now,[30] but I can say to Smith that he is not as alone on this point as he thinks.

* * *

To conclude – a few words about the Hippocratic *Collection*.

According to Edelstein and Smith, it was formed little by little at Alexandria in the course of arrivals of consignments and collections made for the museum. This view of the matter contributes strongly to the skepticism of these authors as far as the whole pre-Alexandrian period is concerned.

I think that the *Collection* was put together, in essence, sooner than that.

That Coan works were studied and used together is a fact that can be deduced from the rich collection of aphorisms entitled *Coan Prenotions* of rather late date, probably the second half of the fourth century (see H. Diller, 1973, pp. 96, 97). There is every chance the *Collection* itself originated in Cos, as the Coan origin of the apocryphal texts would also lead one to believe.

On the other hand, Aristotle's work and the fragments of Diocles of Carystus give proof that a good many of the works of the *Collection* were well known at the height of the fourth century outside of Cos.

The biological works of Aristotle allow for many comparisons with the works of the *Collection*. In this connection, S. Byl has considerably extended and modernized the old and hasty work of Poschenrieder (1887). He patiently demonstrates that more than one hundred passages compare directly with passages from the *Collection* (S. Byl, 1980). These similarities do not give proof, in every case, that Aristotle draws on such and such a work in the *Collection* but there are corroborations that offer strong probabilities and S. Byl (1977, p. 325) thinks that Aristotle had read and made use of more than twenty-five works in the *Collection*.

Certainly one could still believe that Aristotle draws inspiration from works lost to us which treat of identical subjects and which would have to be, in many cases, the common sources for Aristotle and for our *Collection*. To insist on generalizing this thesis brings one to the paradox that Aristotle knew only one work in the *Collection*, the *Nature of Man*, but that he had, on the other hand, access to another collection that parallels ours. I do not think that many specialists are prepared to go that far. The simplest explanation is to accept that, very generally, Aristotle draws his inspiration from the works that we have.

The situation is the same as far as Diocles of Carystus is concerned. We have only fragments of his work but comparison shows us that he must have known and used the *Prognostic, Humors, Epidemics I*, several sections of the *Aphorisms, Nature of the Child, Regimen, Joints, Regimen in Acute Diseases* including the appendix, *Diseases III, Eighth Month Child.*[31] Smith, who is particularly interested in the *Regimen*, thinks that Diocles may very well have read this work (see Smith, 1979, p. 167). But Diocles does not specifically mention the *Regimen* any more than other works so that what would hold true for the *Regimen* must also be true for other works.

Wellmann (1901, p. 64) thought that Diocles was the creator of the first Hippocratic *Collection*, but it is hard to see why it would be Diocles rather than Aristotle.

In any case, that is going too far. The *Collection* goes back more probably to the library of the School of Cos[32] but the works that were brought together there and preserved, for obvious reasons, were well known elsewhere from the time of their publication. Everything indicates even, that Aristotle and Diocles had mainly at their disposal medical works that have been preserved for us.

On this point as well, from the latter third of the fifth century until Alexandrin Baccheios, there has been much more continuity than is admitted in certain skeptical constructions.

Université de Mons et
Université de Bruxelles
Belgium

NOTES

[1] These texts pose difficult problems: cf. alternatively, H. de Ley (1969) where the thesis does not seem to me unassailable.

[2] This does not exclude the possibility that physicians can be specialists in the history of medicine and do great things: cf. the work of M. D. Grmek, Paris.

[3] On *Ancient Medicine*. cf. Deichgraeber (1934, p. 170 passim).

[4] In my edition of *Hippocrate* (1978), I repeated Littré's warning that the *Vision* had not been quoted by any of the ancients (p. 163) having failed to note as Smith does (1979, pp. 152–153) a quotation in the *Commentary on Epidemics II* known by the Arabs that Littré could not have known about.

[5] Smith (1979) writes incidentally about Menon: " . . . whether he idealized Hippocrates because Plato had spoken of him as he did or because the reputation of Hippocrates was *already* considerable in the fourth century", (p. 50). The emphasis on already is mine. In any case, Smith should ask himself why Hippocrates was *already* famous in the fifth century.

[6] This confirmation must be remembered before deciding that nothing is worth trusting in the *Lives* of Hippocrates, cf. Bourgey (1953, p. 83).

[7] On Diocles' dates, cf. Fr. Kudlien (1963) and G. Harig-J Kollesch (1974).

[8] A whole series of medical memos from *Epidemics V* come from Larissa, Chapter 11–25 and date from the second quarter of the fourth century shortly after Hippocrates' death.

[9] A Coan inscription unknown until recently introduces a physician, Hippocrates, son of Thessalus, who lived about 200 B.C. Tradition is strong in Cos. Cf. J. Benedum (1980, p. 39).

[10] This concerns the expression *κατ' ἀνδρογένειαν*, Littré IX, 416, l. 17.

[11] Anomaly also because of its unexpected position at the end of the chapter.

[12] Smith's conviction seems rather recent. In his article (1973) he seems to reject the attribution of the *Regimen* to Hippocrates that Galen once makes and in 1978, speaking about the *Regimen* at the Paris Colloquium he made no reference to this thesis; cf. *Hippocratica*, pp. 439–448.

[13] Probably at the fourth Hippocratic Colloquium at Lausanne in 1981.

[14] I still think that *τὸ ὅλον* means the Universe. W. K. C. Guthrie (1975, p. 460) writes: "I doubt if he (Plato) even intended a *double entendre*, though this is possible."

[15] Smith would have been well advised here to follow Heidel's correction, but he follows Littré and translates this last verb by "know", (p. 47). The comparison of texts is made too by Lloyd (1975, p. 172, n. 5) but much more circumspectly and with no positive conclusion.

[16] Cf. my edition (1967) p. 2; translation slightly altered for the edition CMG, Berlin.

[17] Jouanna thinks so, but I cannot agree with him. Cf. note 13.

[18] I think however, that Plato must know the *Regimen*, but it is through other works that one could prove it: notably *Timaeus*, 98 c–d 2, and even the *Republic*, 406 c. My CMG edition will report on this question.

[19] And this is indeed the case for Smith (1979) and Lloyd (1975).

[20] Everything that can be attempted has been well presented by Lloyd (1975).

[21] Chapter XIX. Detailed comparison by J. Jouanna (1975) pp. 56, 57 and notes.

[22] In VIII, 35 A. (Abas ?) remains a mystery to us; in XX 22 and 25, Philolaos and Philistion were very well known if we are to believe the ancients.

[23] Cf. J. Jouanna (1969) p. 156 and (1975) pp. 58, 59. The mention of Polybus as the son-in-law of Hippocrates is rare in antiquity; this is not a cause for suspicion as Jouanna seems to think: there is no mystery in the fact that physicians are more interested professionally in the quality of the student.

[24] In his recent book, G. E. R. Lloyd (1979) does not doubt the existence of medical centres at Cos and at Cnidos (cf. pp. 30 and 98), but this fact plays no role (and I understand it) in his report on the *Collection*, pp. 15–58 and 146.

[25] A point of view very different from mine (because it is set too exclusively in a rigorously logical framework), but very interesting also for our discussion is that of V. Langholf (1980) pp. 333–346.

[26] This is why, for example, the thesis presented by Jouanna in his report (1977b) pp. 291–312 does not seem to me unassailable.

[27] Cf. *Diseases* II, 73; *Internal Affections* 3; 6; 13; 16; 43, etc. The meaning of τὴν ὤρην is guaranteed by *Diseases of Women*, 63.

[28] The most extreme case is undoubtedly that of Jouanna (1974) p. 16; cf. my article (1978) p. 536.

[29] Cf. Joly (1966) pp. 64–69; complete unanimity of opinion does not however exist cf. Fr. Kudlien (1977) pp. 95–103.

[30] A. Thivel, Cnide et Cos? (Paris, 1981) contains a long chapter on this theme and quite recently a work by V. Di Benedetto (1980) gives an undoubtedly extreme point of view, pp. 97–111.

[31] Cf. Wellmann (1901) p. 64. Wellmann adds still others, less certainly used. It is impossible to include as he does *Nutriment* which comes later, (cf. Joly 1972, pp. 132–137) and *Hebdomades* which comes later still according to J. Mansfeld.

[32] This is also the opinion of H. Diller (1973) p. 99. Some later works were of course added afterwards.

REFERENCES

Bachelard, G.: 1947, *La formation de l'esprit scientifique*, Paris.

Benedum, J. 1980, 'Inscriptions grecques de Cos relatives à des médecins hippocratique et Cos Astypalia', *Hippocratica*, Paris.

Bourgey, Louis: 1953, *Observation et expérience chez les médecins de la Collection hippocratique*, Paris.

Bousquet, Jean: 1956, 'Inscription de Delphes' *Bulletin de Correspondance Hellènique* **80.**

Byl, S.: 1977, 'Les grands traités biologique d'Aristote et la Collection hippocratique', *Corpus Hippocraticum*, Mons.

Byl, S.: 1980, '*Recherches sur les grands traités biologiques d'Aristote: sources écrites et préjugés*', Mémoires de l'Académie de Belgique **64**.

Deichgraeber, K. 1934, *Die Epidemien und das Corpus Hippocraticum*, Berlin.

Deichgraeber, K.: 1971, 2nd ed. with *Nachwort und Nachträge*, Berlin, New York.

Diller, H.: 1973, *Kleine Schriften zur antiken Medizin*, Berlin, New York.

De Ley, H.: 1969, 'De samenstelling van de Pseudo-hippokratische Brievenverzameling en haar plaats in de traditie', *Handelingen XXIII der Koninklijke Zuidnederlandse Maatschappij voor Taal-en Letterkunde en Geschiedenis.*

Di Benedetto, V.: 1980, 'Cos e Cnido', *Hippocratica*, Paris.

Dugand, J. E.: 1977, 'Hippocrate à Thasos et en Grèce du Nord', *Corpus Hippocraticum*, Colloque de Mons 1975, Mons.

Edelstein, Ludwig: 1931, *Περὶ ἀέρων und die Sammlung der hippokratischen Schriften*, Berlin.

Edelstein, Ludwig: 1935, 'Hippokrates', Pauly-Wissowa, *RE*, Suppl. Band VI.

Entralgo, P. Lain: 1970, *La medicina hipocratica*, Madrid.

Grensemann, H.: 1975, *Knidische Medizin, Teil I*, Berlin-New York.

Guthrie, W. K. C.: 1975, *A History of Greek Philosophy IV*.

Hercher, R.: 1871, *Epistolographi Graeci*, Paris.

Hercher, R.: 1914, Re-edited, G. Putzger, Ostern.

Joly, Robert: 1961a, 'La question hippocratique et le témoignage du *Phèdre*', *Revue des Etudes Grecques* **74**, 69–72.

Joly, Robert: 1961b, *Recherches sur le traité pseudo-hippocratique du Régime*, Paris.

Joly, Robert: 1966, *Niveau de la science hippocratique*, Paris.

Joly, Robert (ed.): 1967, Hippocrate, *Régime*, Paris.

Joly, Robert: 1972a, 'Hippocrates of Cos', *Dictionary of Scientific Biography* **6**, New York, 418–431.

Joly, Robert (ed.): 1972b, *Hippocrate* **6**, Paris.

Joly, Robert (ed.) 1978a, *Hippocrate* **13**, Paris.

Joly, Robert: 1978b, 'L'Ecole médicale de Cnide et son évolution', *L'Antiquité Classique* **47**, 528–537.

Joly, Robert: (1980), 'Un peu d'epistémologie historique pour hippocratisants', *Hippocratica*, Actes du Colloque hippocratique de Paris, Paris, 285–297.

Jouanna, J. 1969, 'Sur une nouvelle édition de "La Nature de l'homme" ', *L'Antiquité Classique* **38**, 150–157.

Jouanna, J.: 1974, *Hippocrate. Pour une archéologie de l'école de Cnide*, Paris.

Jouanna, J.: 1975, CMG, I, 1, 3, Berlin.

Jouanna, J.: 1977a, 'La Collection hippocratique et Platon (Phèdre, 269c–272a)', *Revue des Etudes Grecques* **90**, 15–28.

Jouanna, J.: 1977b, 'Le problème de l'unité du traité du Régime des maladies aigües', *Corpus Hippocraticum*, Mons, 291–312.

Kollesch, G. Harig-J.: 1974, 'Diokles von Karystos und die zoologische Systematik', *NTM- Schriftenr. Gesch., Naturw., Technik, Med.*, Leipzig, 24–31.

Kudlein, Fr.: 1963, 'Probleme um Diokles von Karystos', *Sudhoffs Archiv.* **47**, 456–464.

Lloyd, G. E. R.: 1975, 'The Hippocratic Question', *The Classical Quarterly* **25**, 172–191.

Lloyd, G. E. R.: 1979, *Magic, Reason and Experience. Studies in the Origin and Development of Greek Science*, Cambridge.

Littré (Edition): 1861, *Oeuvres Completes el Hippocrate*, Paris.

Langholf, V.: 1980, 'Über die Kompatibilität einiger binärer und quaternärer Theorien im Corpus Hippocraticum', *Hippocratica*, Paris, 333–346.

Poschenrieder, Fr.: 1887, *Die naturwissenschaftlicehn Schriften Aristoteles in ihrem Verhältnis zu den Büchern der hippokratischen Sammlung*, Bamberg.

Smith, W. D.: 1973, 'Galen on Coans and Cnidians', *Bulletin of History of Medicine* **47**, 569–585.

Smith, W. D.: 1979, *The Hippocratic Tradition*, Cornell University Press, Ithaca and London.

Wellmann, M.: 1901, *Die Fragmente der sikelischen Aertze*, Berlin.

JAAP MANSFELD

THE HISTORICAL HIPPOCRATES AND THE ORIGINS OF SCIENTIFIC MEDICINE

COMMENTS ON R. JOLY, "HIPPOCRATES AND THE SCHOOL OF COS"

1. "Between myth and skepticism", or, *medio tutissimus ibis* . . . There are many decent things in Professor Joly's[1] paper. Summarizing and to a certain extent reshuffling the argument, I find that it is concerned with five questions: (1) Is there, in the *Corpus Hippocraticum*, a group of interrelated treatises which, with reasonable confidence, may be ascribed to Hippocrates of Cos himself? (2) Is there, in the *Corpus*, a group of treatises, including those to be attributed to Hippocrates, which may be ascribed to a Coan 'school'? (3) Is there, in the *Corpus*, another group of treatises, to be attributed to another 'school', viz., the Cnidian? (4) What is the origin of the main and early body of the *Corpus* qua collection? (5) Is it acceptable that Hippocratic medicine, i.e., that of the *Corpus* in general and of its Coan section in particular, be considered a *scientific* discipline? With as a corollary: are the Coan and the Cnidian parts significantly different in this respect?

J.'s answer to questions (1), (2), and (3) is affirmative. As to (4), the *Corpus*, he argues, originated in the school of Cos. As to (5), his evaluation of Hippocratic medicine is very critical of what he sees as the 'myth' of its scientific level, the difference between Coans and Cnidians, moreover, alledgedly being insignificant.

The first four points, although not without importance for our evaluation of the *Corpus* as a whole, are mainly literary and biographical. The fifth belongs to a different category, viz., that of the history of science as considered from an epistemological point of view. I shall postpone discussion of this last point, on which I disagree rather clearly with J., to the final section of this commentary. On the other hand, agreement between J. and myself on the literary and biographical issues is substantial, such differences as can be pointed out being minimal when compared with what separates both of us from other people in the field. I do not think, however, that the ways along which we tend to reach our almost similar conclusions are as closely together as the conclusions themselves. To put it differently: I do not think all of J.'s arguments equally good. What is more, his suggestion that one should neither

Michael Ruse (ed.), Nature Animated, 49–76.

accept everything nor reject everything is, of course, unobjectionable as a general principle; but its application is not without hazard, inasmuch as one can only determine what is 'mythical' on the one hand and what 'skeptical' on the other if, some way or other, one has a certain premonition about the way things will look when contemplated from the *via media*.

2. J.'s overview of the external and internal evidence pertaining to the historical Hippocrates and to such treatises as may be attributed to him is virtually complete, and his treatment of this evidence judicious.

The first thing he establishes is important (pp. 31–2). Arguing against Edelstein and others, he proves that, for Plato (in *Protagoras*) as well as for Aristotle (in the *Politics*), Hippocrates is *the* Greek physician.

His discussion (pp. 32–3) of the testimony of Ctesias, however, is not satisfactory. I am prepared to accept that Ctesias the Asclepiad, in the early fourth cent. BCE, read and criticized a passage from *On Joints* (IV 288, 11 f. Littré), but cannot be sure that Ctesias actually said that *On Joints* had been written *by Hippocrates*. Galen, to whose *Commentary* (XVIIIA 931, 5 f. Kühn) on the passage at issue (!) we owe our information, merely says: "people condemn Hippocrates for his attempted reduction of a dislocation of the thigh at the hip, for it would fall out again immediately; the first to do so being Ctesias etc.". To state that this is not sufficient as a basis for the attribution of the surgical treatise to Hippocrates does not amount to skepticism, but constitutes an act of justifiable criticism. Galen may well have supplied the reference *to Hippocrates* himself, since "in his commentaries (he) commonly refers to the author of the treatises he is discussing *as Hippocrates*, even where he knows that the authenticity of the treatise in in doubt".[2] The sin of skepticism would come in if and only if, finding in Galen a *verbatim* quotation from Ctesias saying, e.g., "Hippocrates' method of reducing the thigh is wrong etc.", we would raise an eye-brow and ask how Ctesias could know the work he is criticizing on this point is by Hippocrates. I am not saying that Ctesias did *not* mention the name of Hippocrates, much less that we can be sure he did not, but only that we do not know, and have no means of ascertaining, whether he did or didn't. As long as this *non liquet* constitutes the only realistic verdict, the declaration of authenticity of the surgical treatise based on Galen's information regarding the criticism of Ctesias *and others* is not a valid inference, but a contribution to scholary myth.

In as far as the possible authenticity of works in the *Corpus* is concerned, the case of Ctesias, accordingly, is not different from that of Diocles of Carystus, a contemporary of the older Plato and young Aristotle. J. elsewhere in his paper (p. 44) refers to Diocles in order to remind us that he knew several works from our *Corpus,*[3] not in order to suggest that he attributed any of these to Hippocrates. I agree. Actually, the list of treatises known to Diocles as originally composed by Wellmann[4] conveys an important lesson. For Wellmann, basing these particular inclusions upon a passage in Galen and another one in a late medical doxography known as the Excerpt of Vindicianus,[5] listed two 'Hippocratic' works which were written when Diocles had long been dead.[6] The lesson to be learnt, of course, is that passages in commentators or doxographers which, in a general way, speak of certain doctrines as being at variance with certain others, are to be classified as slippery until further notice. This also holds for those cases where we have reason to believe that the precision of such statements of a more definite nature as are to be found in such a source is spurious. From Galen and from another, late commentary Wellmann lists two passages in which Diocles is actually made to address, in the first 'Hippocrates', in the second the author of *Epid.* I;[7] if acceptable, these citations, or at least the first, would provide the certainty we found lacking in Galen's report about Ctesias. However, it is generally agreed that these 'fragments' of Diocles are not quoted *verbatim*, and the second citation does not address "Hippocrates" anyway. J. therefore is quite right in omitting Diocles' evidence from his overview of attestations of works *by* Hippocrates.

One has to conclude, then, that we know nothing of the attribution of treatises to Hippocrates by these fourth-century medical authors. That Ctesias and Diocles knew and used several pieces from our *Corpus* is beyond doubt. We are not in a position, however, to attribute either the surgical treatise or any other piece they knew to Hippocrates on the basis of such evidence as they provide.

Epidemics I and III (and II–IV–VI) is next. J. affirms (p. 33 f.) that this work is attributable to Hippocrates because it fits the more reliable bits of the ancient biographical evidence as preserved, e.g., in the *Life* of ps. Soranus.[8] He points out that II–IV–VI, which, on generally accepted internal evidence related to archaeological and epigraphical data, are probably to be dated to the early years of the fourth cent. BCE,[9] bring us to Thessaly and even to Larissa in Thessaly, where, according to ps. Soranus and the other *Lives* – whose information on this point is generally accepted –, Hippocrates died and was buried (Tzetzes, who follows the lost *Life* of Soranus, also tells us he lived there). J. also points out that *Epid.* I and III, to be dated not before 410 BCE,[10] bring us to the island of Thasus and the adjacent coastal regions, which are not far from Thessaly. He is, of course, aware that the ancient biographies also propagate the 'myth', or legend, of Hippocrates, but insists that they contain certain elements which it would be unwise to throw out together with the legend, viz., "factual questions about relationships of date or place" (p. 34) or about "relatives" (p. 34). Thus, he points out (pp. 34–5) that Hippocrates' son Thessalus, whose name is absent from the earlier treatises in the *Corpus*, not only appears in the biographies, but also in a Delphic inscription. He also insists that Polybus, whom the *Lives* know as Hippocrates' son-in-law, is known to Aristotle as the author of one of the early pieces in our *Corpus*, viz., *Nature of Man*. He further argues that ps. Soranus[11] in his biography "quotes numerous ancient sources and this precludes the apocryphal myth as the *sole* source of the lives" (p. 34; my italics). Exactly.

This argument can be further strengthened. The information of the *Lives* as to the location of Hippocrates' tomb is confirmed by a four-line epigram, *A.P.* VII 135, said to have been inscribed on it. Theoretically, its authenticity is open to question, because it has not survived on stone. Since it contains no reference whatsoever to the Hippocrates legend (to his philhellenism, for instance) such doubt, however, would be hypercritical. I translate lines 1–2: "Hippocrates, by birth a Coan, sprung from the root of immortal Phoebus, lies here a Thessalian". – "Lies here a Thessalian": this suggests an honorific burial,[12] quite possibly even the conferment of Thessalian citizenship. Note that (just as in the *Lives*) not only the dead man's place of birth is mentioned, but also the fact that he is a descendant of Apollo, i.e., an Asclepiad.[13]

In ps. Soranus himself, we have two items, concerned with Hippocrates' dates, which deserve to be closely studied – if only because, wholly or in part, they have been doubted. J., who speaks of "questions . . . of date" in general terms, does not enter into the details.

The passage runs as follows: "His *floruit* is in the Peloponnesian times [i.e., the Pelop. war], and he was born in the first year of the 80th Olympiad [460/59 BCE], as Ischomachus[14] says in the first book of his *On the Sect of Hippocrates*; and, as Soranus of Cos,[14] who made investigations in the Archives at Cos, adds, when Abriadas was *monarchos*, on the 27th day of the month Agrianios, which is why, he says, the Coans up till now sacrifice to Hippocrates on this day".[15]

I shall discuss the evidence of Ischomachus first. His terminology reveals that what he gives us is the ancient chronographical vulgate[16] of Eratosthenes and Apollodorus, after the prose abstract from the latter which converted Apollodorus' Athenian archon years into Olympiads.[17] One of the mainstays of Apollodorus' chronological system is the *floruit* of a person, i.e., the 40th year of his life. If he dated Hippocrates' birth to 460/59 BCE, the *floruit* – reckoning inclusively – is 421/0 BCE, which indeed provides a rough synchronism with the 'times' of the Peloponnesian war. Apollodorus synchronized political and cultural history; 421/0 BCE is the year following upon the first phase of the war (peace of Nicias March 421, i.e., 422/1 BCE). I conclude that Ischomachus' (hence ps. Soranus') ultimate source for these dates is Apollodorus. There is a difficulty here, however, inasmuch as ps. Soranus, previously, mentions Apollodorus [*FGrH* 244 F 73a] – together with Eratosthenes, Pherecydes,[18] and Arius of Tarsus[19] – for the *genealogy* of Hippocrates only. But this difficulty is not unsurmountable. Ps. Soranus' source for the genealogy is not the same as that for the vulgate chronology, which omitted to mention Apollodorus' name. Likely enough, it was Arius of Tarsus who cited "Eratosthenes, Pherecydes, and Apollodorus"; presumably, he, too, only used an abstract from Apollodorus, which preserved the latter's references to Eratosthenes (who may have cited Pherecydes in the first place) and Pherecydes. Apollodorus loved to adduce a variety of older sources.[20] For the genealogy, then, ps. Soranus cited a late source which cited an earlier source which cited even earlier sources. That the chronology given by Ischomachus, which reached ps. Soranus by a different route,[21] is indeed Apollodorean is confirmed by another item in the *Life*. According to "some" authorities (κατὰ . . . τινας), ps. Soranus says (after his genealogical piece and mention of Hippocrates' real teachers and before his reference to Ischomachus), Hippocrates was not the pupil of his father Heraclides and then of Herodicus, but of Gorgias and Democritus. First, it should be noted that Heraclides also occurs in ps. Soranus' Apollodorean genealogy, and is mentioned as one of Hippocrates' teachers in the sentence that follows; secondly, that Apollodorus generally said who begot, and especially who taught,[22] whom: this, of course, was essential to his chronological system. Thirdly, although Apollodorus' dates for Gorgias cannot be established with sufficient precision from the other evidence we possess (*FGrH* 244 F 33), his date for Democritus is known. Apollodorus held that Democritus was born in 460/59 BCE [*FGrH* 244 F 36 = *Vorsokr.* 68 B5, Diog. Laert. IX 4], which is the year of birth of Hippocrates according to the Apollodorean vulgate as preserved by Ischomachus. In Apollodorus' chronological system, pupils are generally 15 or even 40 years younger than their teachers. If both Hippocrates and Democritus are born in 460/59 BCE, a teacher-pupil relationship in conformity with the Apollodorean canon is impossible. The rejection of the heterodox opinion which made Hippocrates the pupil of Democritus, as found in ps. Soranus,[23] is therefore consistent with the date of birth preferred by ps. Soranus, which is that of the vulgate. Furthermore, Apollodorus generally named the authorities he followed and referred to those he disagreed with without revealing their names.[24] I therefore believe that the whole of ps. Soranus §§ 1–2, p. 175, 3–9a Ilberg, *FGrH* 244 F 73a, is a 'fragment' of Apollodorus which reached the *Life* by a route *a*, and that § 3a, p. 175, 9b–11a Ilberg, also is a 'fragment' of Apollodorus, which reached the *Life* by a route *b*. These 'fragments' derive from what, in Apollodorus, originally was a piece of his *bios* of Hippocrates.[25] It is, of course, also possible that ps. Soranus had one source for both the genealogy and the Apollodorean chronology, and that he substituted his posh source for the vulgar vulgate. Whatever the truth, what we have here is the ancient chronographical vulgate, to which we had better stick as long as no other data are available.[26] Jacoby's (later) pessimism is unnecessary anyway.[27]

In this connection, it is important to observe that the dates as provided by Ischomachus are independent of the information offered by Soranus of Cos, said by ps. Soranus, *expressis verbis*, to be an *addition* to Ischomachus.[28] Soranus of Cos refers to another, viz., a *Coan*, chronographic system: he gives the year according to the *monarchos*, i.e., the *Coan* eponymous magistrate, and adds a month from the *Coan* calendar as well as a day in this month. These data, it is said, he discovered by investigating the *Coan* archives. What he also did is that he *synchronized* the year of Abriadas with the Olympiad year 80.1 of the Apollodorean vulgate; at any rate, this is what ps. Soranus implies.

The evidence Soranus of Cos claimed to have discovered has been judged unhelpful,[29] or even faked.[30] Dating by eponymous magistrate, month, and day is believed to be a Hellenistic practice which cannot have been observed in the fifth cent. BCE. Furthermore, evidence regarding the Coan eponymous *monarchos* is limited to the period after 366 BCE, the year in which the new city of Cos was synoecized on the north shore of the island, and (still) is lacking for the years immediately subsequent to the synoecism.

To take the second of these points first: the title *monarchos* for the yearly eponymous magistrate is limited to Cos. It is arguable that it must predate the synoecism of 366 BCE, because it reflects the change from hereditary kingship to yearly office (cf. the Athenian *archon basileus*). Therefore, it is more likely than not that *monarchos* was already the title of the yearly eponymous magistrate of the earlier capital, Astypalaea.[31] It should be added that the name Abriadas, according to Pape-Benseler, *Wörterbuch der griechischen Eigennamen*, and also to the onomasticon of Coan names compiled by Ms Sherwin-White,[32] is a *hapax*, i.e., only occurs ps. Soranus, *loc.*

cit. Like 'Asclepiades', it is an archaic type of name, in *-ades* or *-adas*.[33] The chances that a forger will invent a unique name instead of simply using a more familiar one are minimal; also note that Soranus of Cos uses a typical "archon formula" (*μοναρχοῦντος* 'Αβριάδα), and that the genitive form of the name is in Doric, not in *koine* Greek, which agrees with the claim that he cites what he found in the Archives of Cos, a Dorian city.

As far as I am aware, it is generally assumed (e.g., still also by Ms Sherwin-White) that what he claimed to have found there was evidence contained in a public record office containing birth registers. Such a register, it is generally agreed, if from the fifth cent. BCE, cannot have used a dating system involving eponymous magistrate, month and day. Perhaps this argument is not fully cogent: if Soranus really consulted a birth register, this may have contained the files newly established at the time of the synoecism, which, according to a majority of scholars, may also have been the date of the first eponymous *monarchos*. Since such files contained the evidence upon which claims for rights of citizenship were based, they must have contained information about the parents and grand-parents of the generation of Coans which participated in the synoecism, and they may quite well have adopted the new dating-system – if this was introduced at the time. All this, however, is rather too hypothetical, and such assumptions are unnecessary anyway, since the inference that Soranus claimed to have consulted a birth file is false.

What Soranus tells us is that the Coans, up to his own times, sacrified in commemoration of the late Hippocrates on the 27th of Agrianios, on which day Hippocrates was born in the year of Abriadas.[34] Edelstein argued that this refers to the heroization of Hippocrates, which he dates to late Hellenistic times.[35] The institution at issue, however, is the *hemera eponumos* or 'name-day', already attested in Herodotus.[36] That *enagismata*, "funeral offerings" (cf. Soranus' *ἐναγίζειν*) were offered to dead persons on their *birth-day* is attested in Epicurus' will as preserved, *verbatim*, in Diogenes Laertius. Epicurus here provides for moneys to be spent "for the *enagismata* to my father and my mother and my brothers and myself – in view of the customary celebration of my birthday on Gamelion 10th each year" (Diog. Laert. X 18). Boyancé, commenting on this passage, says: "Il ne s'agit donc ni d'une apothéose, ni même d'une héroïsation, au sens de ce mot qui le rapproche de l'apothéose".[37] At first blush, this act of traditional piety is surprising on the part of Epicurus, who did not believe in (personal) survival after death. We know, however, that he carefully observed the existing religious customs[38] – which, again, proves that this offering of *enagismata* to the, not-necessarily heroic, dead actually was such a custom. Now the celebration of the birth-day of the late Epicurus[39] was a private affair, or rather the affair of a small coterie. According to Soranus, on the other hand, the celebration of the birthday of the late Hippocrates was a public affair, since 'the Coans' (i.e., the Coan state or community), not just a group of Coans, observed it. In other words, the birth-day of Hippocrates was an official festival day in the Coan calendar. Such official festival days, again, were instituted through a decree passed by a legislative assembly.[40] The pieces of the puzzle now begin to fit: what Soranus claimed to have found in the Coan Archives was not a birth register, but a decree, voted by the Coan People, concerned with the festival of Hippocrates, which stipulated the day on which it was henceforward to be celebrated, i.e., which said "The People etc. etc., acknowledging that Hippocrates was a benefactor of etc. etc., in view of the fact that he was born in the year of Abriadas on Agrianios 27th, decide: that on the day afore-mentioned, the People of Cos shall sacrifice *enagismata* in commemoration of Hippocrates etc. etc.", or words to that effect. The Archive was that containing official state documents.[41]

This implies that the Coan authorities, or whoever drafted the decree that was voted, are the 'source' for the date mentioned therein (if thought wrong, it would have been amended by the assembly; if correct, people voting in favour of the decree had reason to believe the date was correct). This, again, implies that 'sources' were available to these people for the date concerned; and I would suggest that the 'source' from which they derived the year was not the same as that from which they got the month and day. The source for the year must have been a list of eponymous *monarchoi* which must have contained names predating the synoecism,[42] in fact by as much as a century and more. Lists such as these probably were only composed from the late 5th cent. BCE onwards; but this does not permit us to assume that the data from earlier times contained therein are spurious.[43] The year itself, viz., the year of Hippocrates' birth, may have been *computed* by reckoning backwards from the year of death, if a memory of the year of death and of the age at which Hippocrates died had indeed survived. This was given according to the Coan list of *monarchoi*; had Hippocrates been an Athenian, the year would have been given according to the Athenian *archon* list. I suggest that the Coans' source of information for year of death and age of Hippocrates in that year were the Asclepiads, who also will have provided information about his day of birth. In fact, this day may even have been celebrated by the *koinon*, or society, of the Asclepiads long before it became an official Coan festival.[44]

That a memory of deceased members of the *koinon*, at least in as far as they were true-blue Asclepiads, was preserved by the associated physicians themselves is proved by the already famous inscription from Delphi, *t.a.q.* 360 BCE.[45] This contains a 'decree' of the association of Asclepiads both of Kos and of Cnidus,[46] stipulating that "any Asclepiad who comes to Delphi, if he desires to make use of the oracle or to perform a sacrifice, has to swear [sc., to the Delphians] that he is an Asclepiad by male descent";[47] a lacuna follows, of several lines, but from the sequel it appears that Asclepiads by male descent, if they swear this oath, enjoy certain prerogatives. This inscription is contemporary with Hippocrates' old age, at least not much later than his presumable year of death. Now, in order to be in a position to know who were Asclepiads by male descent and who were not, the *koinon* must have had a file registering births: who begat whom, etc., and this they must have kept up to date and preserved. [This file the Delphians had not, and therefore the oath was necessary]. In this way, they knew which members were true-blue Asclepiads[48] and which were not. In this context, it is important to observe that Hippocrates is the earliest Greek physician of whom we know that he taught for a fee,[49] i.e., admitted pupils from outside his family, or clan.

Confirmation that the Asclepiads indeed preserved this sort of information turns up in an unexpected place. According to Tacitus, *Ann.* XII 61, the learned Roman emperor Claudius, in the year 53 CE, put a proposal before the Senate to grant the Coans immunity from taxation. Having said something about the (legendary) ancient history of the Coans, he went on to say that in the island "the art of healing was introduced by the advent of Asclepius, and became absolutely famous among his descendants [: the Asclepiads], *citing the names of individual persons and the times at which they had lived*".[50] How bored the Senators must have been! He added that Xenophon, his own *Leibarzt*, belonged to the same family [: was a Asclepiad], and that Xenophon's request for immunity for the sacred island should be granted. Tacitus, with his habitual sarcasm, comments that numerous benefits conferred by the Coans upon the Romans could have been adduced, but that Claudius omitted these, because his thought had only been of Xenophon (i.e., of himself). We know this Gaius Stertinius Xenophon quite well from inscriptions recording the eminence to which he rose in the island itself.[51] It is a reasonable assumption that it was Xenophon the Asclepiad who provided the emperor with at least part of the material used in the latter's speech before the Senate, even though we need not assume that no other sources, dealing, among other things, with the chronology of the Asclepiads (*quibus quisque aetatibus viguissent*), were available to Claudius, whose speech predates the *t.p.q.* of ps. Soranus by a mere few decades.

In conclusion, I would say that we have no valid reason to reject the dates of ps. Soranus, viz., the chronographical vulgate cited from Ischomachus, as synchronized with a date of birth provided by Soranus of Cos. We may, therefore, return to J.'s argument concerning the attribution of *Epid.* I and III to Hippocrates. I, for one, agree with his verdict that such of the biographical data as pertain to Hippocrates' family, year of birth, and place of death, can be accepted.

Ps. Soranus and the other *Lives*, confirmed by the epigram, say that Hippocrates' tomb was in Thessaly. It should be noted, however, that not *Epid.* I and III, but only the somewhat later II–IV–VI, which convey the impression of rough drafts left unfinished at the author's death,[52] are connected with Thessaly. There is nothing in the *Lives* which relates to the topography of I and III. It follows that we can hardly use them (or the epigram) in order to justify, *pace* J., the attribution of *Epid.* I and III (not II–IV–VI!) *to Hippocrates.*[53] If, on the other hand, one accepts that I and III and II–IV–VI, although there may be a certain number of later additions,[54] were all written by the same person – which I see no reason to doubt –, the argument that is valid for II–IV–VI, by implication, is also valid for I and III.

For all that, this evidence is not, as yet, sufficiently compelling to warrant the attribution of *Epid.* I and III + II–IV–VI *to Hippocrates*. All J. may be said to have proved is that the biography and chronological data are *compatible with* the attribution to Hippocrates. This conclusion is important enough. It has not been proved, however, that the attribution is *entailed* by the chronographical and biographical data.

3. The next stage in J.'s argument deals with Wesley D. Smith's recent suggestion that *Regimen* should be attributed to Hippocrates (pp. 35–8). J. offers a refutation; in my opinion, he is rather successful. One of his point, viz., that Hippocrates, if he wrote *Regimen*, cannot also be the author of *On Joints* and *Epid.* I and III, though correct, does not entail that he did *not* write *Regimen*, because, if I have argued correctly in § 2 of this paper, J.'s arguments in favour of attributing *On Joints* and *Epid.* I and III are too weak. A point which he refrains from bringing up is that the probable date[55] of *Regimen* is incompatible with the chronology of Hippocrates' life; this is because J. believes that it was written ca. 400 BCE.[56]

The other parts of his critique of Smith are cogent. Smith's argument rests upon an appeal to Meno's report about Hippocrates' aetiology of diseases in the *Anonymus Londinensis*, so-called, and to Plato's descriptive analysis of Hippocrates' scientific method in *Phaedrus*. As J. points out, Meno's aetiology, found by Smith in a passage of a chapter of *Regimen*, is absolutely irrelevant to the treatise as a whole. Smith also found an explicit reference to a method of collection and division (see Plato, *Phaedr.* 270 a ff.) in another chapter of *Regimen*, but the text, as J. points out correctly, does not bear out this interpretation, and he rightly insists that, according to Plato, the dialectical method is *the* method of Hippocrates, hence should be *the* method used in *Regimen*, which it is not – just as Meno's aetiology is not its aetiology. Now, Plato also said that Hippocrates used this method "not without the study of the nature of the whole". Smith does not attempt to choose between the rival interpretations of the words "the whole" which have been proposed by scholars, viz., the 'whole of body' *vs* the 'whole of nature'. J., quite rightly, I believe, argues that the second of these interpretations is to be preferred. On the other hand, one should acknowledge that Smith successfully quotes a passage in *Regimen* (I 2, beginning with the sentence where he thinks the dialectical method is explicitly referred to), in which a sort of 'division' of the body in relation to a study of the whole of nature is presented as what Smith calls the "impressive outline of a science".[57] J. is silent about the problems Smith's insistence upon this chapter raises for his own interpretation, and rests content with a reference (p. 36) to his own paper of 1961 about 'meteorologia' and the study of the whole. For several reasons, this is to be regretted. The first, which perhaps is only of minor importance, is that, if Smith's case has not been presented in an adequate way, its refutation suffers, too.[58] A more important reason for regret, however, is that in this way J. prevents himself from paying the same judicious attention to the testimonies of Meno and Plato that he has given to those of Ctesias and ps. Soranus. My own view, which I proposed in the original version of this commentary as read at Montreal,[59] is – briefly – as follows.

Plato and Meno (in that order) are our earliest witnesses for *Hippocrates*' ideas. However, the Meno-section on Hippocrates in *Anon. Lond.* can hardly be said to represent in a fair way what Meno must have written originally. *Anon.* contains abstracts from Meno made by a person familiar with the tenets of Stoicism. We should not, therefore, take him as our starting-point or use him on a par with Plato. So only the passage in *Phaedrus*, 269 e ff., is left. A new reading of what Plato says gives us the key to the *Corpus*: Plato must be thinking of *Airs Waters Places*. The dominant and novel theory of this famous little work consists of a typology of human bodies as conditioned by the various types of natural environments inhabited by human populations; it is this theory which is described by Plato as a division, in respect of their active and passive capacities, of bodies in relation to the study of "the whole".

Airs Waters Places thus is to be ascribed to "Hippocrates the Asclepiad". The next stage is obvious: it is relatively easy to add other works, such as, e.g., *Epid.* I and III, once this Archimedean *dos moi pou sto* has been found. Here, I have reasons to assume, Professor Joly and the present writer part company; Professor Smith (*p. litt.*) appeared to disagree as well.[59a] Until my interpretation of *Phaedr.* 269 e ff. has been refuted, however, I shall hold on to my conviction that *Regimen* remains an unlikely proposition, and that Plato's evidence, if properly understood, far outweighs that of Ctesias *e tutti quanti*. Moreover, Plato's evidence, which allows us to authenticate *Airs Waters Places* and, e.g., *Epid.* I and III, is fully compatible with the chronographical vulgate of Ischomachus. *Airs Waters Places* is a fine archaic piece which may be dated to the time at which Ischomachus puts Hippocrates' *floruit*. *Epid.* I and III, which exhibit a further refinement of the environmental theory of *Airs Waters Places*, are to be dated (in view of internal references confirmed by archaeological data) to *t.p.q.* 410 BCE,[60] i.e., about 10 to 15 years later than *Airs Waters Places*. The personal and doctrinal development of the author as traceable in these works is thus consistent with the chronological evidence. The notes[61] posthumously collected as *Epid.* II–IV–VI, then, would give us the final stage of the development of the author of *Epid.* I and III; consistency again, for Hippocrates died, and was buried, in the regions which were the scene of the activities of this author's final years. When we compare the early *Airs Waters Places* to the much later *Epid.* II–IV–VI as to their doctrinal contents, we find that the continuity of general theory far outweighs such differences[62] as can be pointed out. The discrepancies themselves can be fully accounted for, once it is realized that the chronological gap separating the late from the early work is from 25 to 40 years, which is even more (or at least not less) than the difference, in years, between Plato's *Phaedo* and *Timaeus* or *Theaetetus*. J. is right when he insists (pp. 40–1, 42) that Hippocrates' thought will have evolved. I should add that neither J. nor the present writer argue from doctrinal differences to chronological assumptions; quite the opposite: the independent chronological data explain the differences in doctrine.

4. I can be brief about what J. (pp. 39–43) says regarding the attribution of treatises in the *Corpus* to a Coan and a Cnidian 'school', respectively, for I do not only quite agree, but also find his *mise à point* quite good – especially regarding the Cnidian works. As to his argument concerned with a group of Coan treatises, I would say that, once it is accepted that *Airs Waters Places* and *Epid.* I and III + II–IV–VI are by Hippocrates, *Epid.* V and VII and

Humours must be called Coan, and so must *Prognosticon. Nature of Man* is Coan because it is by Hippocrates' son-in-law Polybus. Other treatises can thus be added.

One is also grateful for his reminder that a number of treatises from the *Corpus* were already known to Ctesias (pp. 32–3), Diocles (p. 44), and Aristotle (pp. 39, 43–4). J. assumes that, already in the fourth cent. BCE, there existed a *Corpus Hippocraticum*, and he accepts Diller's[63] argument that this collection originated in the Coan 'school', viz., was their 'library' (p. 44). I believe that Diller's and Joly's suggestion can be further strengthened.

The *Corpus Hippocraticum* we have is unique among the great ancient literary corpora. Even if we deduct those works which adhered to it in Hellenistic and Graeco-Roman times, and concentrate on the majority of the treatises it contains and which date from the fifth-fourth cent. BCE, we still are faced with a rather motley collection: Coan works, Cnidian works, and individual pieces such as *Ancient Medicine, On the Art*, and *Regimen*. The case of two corpora, of similar size, of two other early authors is different. There is no doubt whatever that the great majority of writings contained in the *Corpus Platonicum* is indeed by Plato, as there is no doubt either that the great majority of writings in the *Corpus Aristotelicum* is indeed by Aristotle. Here the *Corpus Hippocraticum* differs rather sharply; however, a comparison of these three corpora is also useful in another way. As far as its authentic ingredients are concerned, our *Corpus Platonicum* consists of pieces written for general circulation, indeed published by Plato himself (or, as in the case of the *Laws*, written for publication and published by Plato's literary executor soon after his demise.)[64] The genuine parts of our *Corpus Aristotelicum*, however, as everyone knows are exclusively works from Aristotle's *Nachlass*, which, presumably, were not intended for publication by their author. The works he published himself are lost. The *Corpus Hippocraticum*, on the other hand, not only comprises works which, as to their character or genre, are comparable to the genuine parts of the *Corpus Platonicum* and a number of lost works by Aristotle, but also other works, comparable to the authentic pieces in Aristotle's *Nachlass*. To give a few examples: *Diseases* I, a rather late Cnidian work, explicitly addresses itself to the general public; *On the Art* defends the status of medicine as a scientific discipline, which only makes sense if it was a public defence against an equally public attack.[65] Plato not only had read *Airs Waters Places* himself, but also, in *Phaedrus*, can have one of its *dramatis personae* refer to it in a way which shows that it was widely known.[66] The several groups of books constituting the *Epidemics*, on the other hand, cannot have been written with an eye to publication. This is not only apparent from the succint and rough manner in which they have been composed, and from the fact that the unfinished drafts constituting II–IV–VI were not even rewritten to make them conform to the more finished form of I and III, but also from the fact that they very often mention the patients, whose ailments were recorded, *by name*, which precludes that their author intended them to be published. *Epid.* I and III + II–IV–VI constitute a *Nachlass* which continued to be used, and V and VII constitute additions made to the *Nachlass* in the decades after Hippocrates' death.[67] An obvious *analogon*, from the *Corpus Aristotelicum*, is the *Historia Animalium*, of which scholars have argued that books I–VI and VIII are by Aristotle, although additions, they say, were made by his pupils; VII is of at least doubtful authenticity, and IX a Peripatetic work.[68] To put it differently: *Airs Waters Places*, in a way, is Hippocrates' *Peri Philosophias, Epid.* I–VI his *Metaphysics* (*mutatis mutandis*, of course). The *Epidemics*, anyhow, were the private property of a small association of physicians (just as the *Hist. An.* were the property of the Peripatetic Society), who went on collecting, and adding, material.[69] Thus, the *Epidemics* alone are evidence proving the existence of a working library of sorts in a Coan 'school' around the mid-fourth cent. BCE. A library which, at the very least, also contained the *Prognosticon*, for H. Grensemann[70] has proved, in an examplary case-study, that the observation of patients by the author(s) of *Epid.* V and VII was conducted along lines prescribed in the *Prognosticon*.

Did this library also contain other works? The author of *Epid.* III (Hippocrates, I presume), at the end of the collection of general remarks contained in this book, observes (Chapter XVI): "The power, too, to study correctly what has been written [τὰ γεγραμμένα] I consider to be an important part of the art of medicine.[71] The man who has learnt these things and uses them will not, I think, make great mistakes in the art" (tr. Jones). "What has been written", sc., about medical matters: we should not, I suggest, think of the *Prognosticon*, or even of *Epid.* I and III itself, at least not exclusively, but accept τὰ γεγραμμένα in the large and general sense these words convey. "What has been written" pertains to the *medical literature in general*. This brings us in a position to solve two problems at one blow, viz., (1) why the *Corpus Hippocraticum* contains works written by a plurality of authors and why in it the authentic works of Hippocrates probably only form a substantial minority; and (2), why all these works came to be attributed *to Hippocrates*. The author of *Epid.* III believes that a critical *study* of the medical literature is indispensable to whoever wants to be a successful practitioner of the art of medicine. For this purpose, he (Hippocrates) collects a number of important works written by various professionals other than himself, and studies them. Together with the books written by himself, these works were inherited by his successors, who, moreover, took the injunction of *Epid.* III, Chapter XVI, to heart and kept on adding to the collection [not only, as we have noticed, by writting certain things themselves, but also] by acquiring other people's works. In how far, or for how long, a knowledge of who originally had written what survived, we cannot know, although traditions concerned with the composition and growth of the *Epidemics*, as reported by Galen, have been confirmed by the modern analysis of this work, carried out without recourse to these traditions.[72] The *Coan Prenotions*, on the other hand, a collection of systematic excerpts to be dated, according to Diller,[73] to *t.p.q.* 350 BCE (but the *t.p.q.* will be later), are made up not only of abstracts from Coan, but also from Cnidian treatises,[74] which shows that, at the time of composition of the *Coan Prenotions*, these already belonged to the 'working library'. Diller, again, has pointed out that the presence of a substantial body of Cnidian writings in the *Corpus* is only explained when it is assumed that "die koische Ärzteschule ... auch die Werke der anderen Schulen in ihrer Bibliothek aufgenommen hatte".[75] We should not, however, speak of a Coan school of medicine as distinguished from a Cnidian school of medicine, for Coans and Cnidians, after all, formed one *koinon,*[76] but of the 'school' *of Hippocrates* which not only preserved the writings of the great Hippocrates, but also preserved and collected other important writings, continuing a tradition which, if my interpretation of *Epid.* III Chapter XVI is correct, was begun by the master himself. It is this collection which passed to the Alexandrians.[77] Now we know that the history of the discussion concerning the authenticity of writings by Hippocrates also begins in Hellenistic times.[78] That a corpus was nevertheless accepted as being by Hippocrates shows that it was transmitted to the Alexandrians as 'the books of Hippocrates'. The edition of the *Corpus Platonicum* by Aristophanes of Byzantium constitutes an obvious parallel; Müller has shown that the fact that Aristophanes' edition contained the (spurious) *Epinomis* and *Minos* among the genuine works of Plato proves that the collection of books by Plato transmitted to him was 'beglaubigt', i.e., was the oeuvre of Plato as 'edited' by the Academy.[79]

The suggestion that the earliest part of our *Corpus Hippocraticum* was (written and) collected by none other than Hippocrates himself, who was merely imitated by his successors in this respect, is, of course, speculative. But it accounts, at one blow, for two otherwise irreconcilable facts, viz., the variety of the contents of the *Corpus* on the one hand and the unanimity of their official attribution on the other.

Aristotle (still) knew that *Nature of Man* is by Polybus, but this medical manifesto may quite well have circulated separately. Echoes of individual treatises, in Ctesias, Diocles, or Aristotle,[80] may reflect the circulation or at least availability of individual pieces, especially in medical circles; J.'s playful suggestion (p. 44) that Aristotle is a possible candidate for the honour of being the *Corpus*' first collector is not as wayward as it seems, for Aristotle may have formed a collection of his own, preserved in the library of the Peripatus. Meno, Aristotle's pupil, will hardly have written a *History of Medicine* without a medical library (to judge from the abstracts in *Anon. Lond.*, this also contained works outside the *Corpus*). J. argues, however, that the echoes in Diocles and Aristotle should be explained by the assumption that these knew the [early portion of] the *Corpus* as we still have it; but this argument is tenuous, since we cannot detect echoes, either in the *Corpus Aristotelicum* or in the fragments of Diocles, of works that have been lost. The suggestion, on the other hand, that the 'Library of Hippocrates', the greatest of all Greek physicians according to Plato and Aristotle,[81] may have enjoyed a certain prestige is not implausible. If this is correct, it may, moreover, have been accessible to others: to other doctors and to serious scholars, for instance, who even, if they so wished, could have copies made, or who were permitted to make abstracts. It should be pointed out that *Hist. An.* X, both spurious and post-Aristotelian (it did not, apparently, always figure as the tenth book, but also existed as a separate treatise) is largely made up of extracts from the gynaecological treatises of the *Corpus Hippocraticum.*[82] The assumption that these specialist works circulated freely is hardly feasible, and the fact that such abstracts were made by Peripatetic scholars and thought worth preserving confirms this impression; their inclusion in the *Hist. An.* is valuable from another point of view, viz., as confirmation of the mechanism at work when the attribution of works contained in or connected with a *Nachlass* (which continued to be used) is at issue.

5. In the first part of his paper (p. 29), J. criticizes what he calls "the valorization of Hippocratic medicine, of the Collection in general, and of the School of Cos in particular", which consists in the tenet that "Hippocratic medecine is conceived of as very close to contemporary medecine" (p. 29). For the argument at issue,[83] we are referred (p. 30) to a well-known book (Joly, 1966) and to a recent paper (Joly, 1980). Near the end of his present paper, he returns to this problem; although he accepts that Cnidians should be distinguished from Coans, he rejects the modern tendency to "radically

separate their doctrines into concepts of medicine belonging to Cos or to Cnidus, often in the desire to elevate Hippocrates and his school in relation to an archaic, mediocre, bumbling Cnidian school" (p. 42). He goes on to argue that, for the last fifteen years or so, he has been convinced that such differences between Coans and Cnidians as exist are relatively unimportant, and that Wesley D. Smith, who argued *contra* the distinction of Coan and Cnidian medical 'schools',[84] "is not as alone on this point as he thinks" (p. 43).

I have always admired the originality of Smith's paper, so naturally did check J.'s references to his own publications. In 1966,[85] all he said is that Coan and Cnidian *gynaecological* ideas are almost identical [this, if true, which is not at issue here, would be puzzling] and that both have the same "mentalité scientifique, *sur ce point au moins*".[86] In 1966, then, there is no anticipation, at any rate only one of a very restricted sort, of Smith's argument. It has to be granted, however, that in 1972, in the introduction to vol. VII of the Budé *Hippocrate* and in an article (Joly, 1972), he goes much farther, and that his views, though not as radical or as consistent as those published by Smith in 1973, are to a certain extent compatible with the latter. Yet I note that, in the article, J. speaks of the "Coan doctrine of ambient factors",[87] and also stresses the fact that, unlike *Airs Waters Places*, no Cnidian treatise presents a 'systematic exposition of' these factors.[88] The present writer finds J.'s position (Joly, 1972) much more acceptable than that of Smith. Indeed, Smith, and J. in so far as, to-day, he follows Smith have swopped the learned myth of the fundamental distinction between *opposed* schools (Cos *vs* Cnidus) for a position I can only see as hypercritical. I still believe that an important difference between Coans and Cnidians can be pointed out, important not only from a historical, but also from an epistemological point of view. Although both 'schools' may be called 'scientific', they are so in (partly) different ways.

In his *Phaedrus*, Plato mentions the distinguished Coan Hippocrates by name and clearly identifies him for us. He ascribes to him a classificatory typology of human bodies related to the classifiable type of environment people live in.[89] He does *not* ascribe to him a division and collection, i.e., a classification, of diseases. According to Plato, Hippocrates, as a scientist, is interested in a sort of systematic naturalist anthropology, not in systematic nosology. Cnidian medicine, on the other hand, especially in its earliest form as known to us, is first and foremost concerned with classificatory nosology. The lost *Cnidian Gnomai* are criticized by the author (not necessarily Hippocrates, I think) of *Regimen in Acute Diseases,*[90] and Galen has preserved a fragment from a work of the early Cnidian physician Euryphon, which, apart from some textual variants, is identical with Chapter 68 of the oldest Cnidian treatise we possess, *Diseases* II A.[91] From the criticism formulated in *Regimen in Acute Diseases* as well as from the practice observed in *Diseases* II A, it can be established beyond doubt that the Cnidians set out sophisticated classifications of diseases, not, as Hippocrates, of environments and human groups.[92] It should be granted, naturally, that later Cnidian works exhibit a certain interest in the environmental factors which are crucial to *Airs Waters Places* and *Epid*. I and III. However, if we think of early (and of the dominant worries of later) Cnidian medicine only, we cannot but acknowledge that there is a capital difference between its nosological classification and the environmental theory and human typology of *Airs Waters Places*. In *Phaedrus*, Plato, discussing dialectic, esp. the art of division, as the proper method of science, could never have instanced early Cnidian nosology, because the early Cnidian method of classification is (1) arbitrary in detail and reckless in making distinctions or acknowledging similarities, and (2) not related to 'the whole'.[93]

Consequently, when we try to distinguish Coans from Cnidians, we should not attempt to distinguish *the* Coans from *the* Cnidians, let alone a Coan 'school' from a rival Cnidian 'school'; what we really ought to do, if only for starters, is to distinguish, whenever possible, between individuals:[94] the Coan physicians Hippocrates and Polybus on the one hand, Cnidians such as the authors of the *Cnidian Gnomai*, and Euryphon (if the latter is not, as some say, the author of the *Cn. Gnomai*)[95] on the other.

Now one of the reasons for J.'s reluctance to distinguish between Coans and Cnidians, it will be recalled, is his conviction (see Joly, 1966) that Hippocratic medicine in general is pre-scientific, or merely 'rational', because the Hippocratic physicians were the victims of 'obstacles épistémologiques' of all sorts. He is, of course, right that Hippocratic medicine cannot compare with the medical science(s) of our own time. To say that, for that reason, it should be dubbed pre-scientific, however, goes too far, and the notion of the epistemic obstacles contributes an obstacle impeding impartial evaluation.[96]

First, we should acknowledge what the ancients themselves have to say. The epigram said to have been inscribed on Hippocrates' tomb, lines 3–4, states that he gained many victories over diseases, using the arms of Health, and did so "not by accident, but through his *techne*".[97] Plato, *Phaedr*. 269 e ff., is undoubtedly justified in pointing out that 'great' medicine, e.g., that of Hippocrates, is a scientific *techne* because it has a *method*, i.e., uses a sort of *dialectica utens*, and has extracted from *meteorologia* ('natural philosophy') "what was suitable to the existing discipline [*techne*]".[98] Hippocratic medicine, according to Plato, is an empirical science[99] which uses a scientific method and works out a suitable medical 'physics'. In this respect, I would say, Hippocratic medicine is not, in principle, different from the medicine of our own times. Our physics, chemistry, biology and technology are, of course, far superior. But Hippocrates could only use what was available in his own day.

However, it is also arguable that the early Cnidians, too, although in a far more modest way than Hippocrates, abstracted from natural philosophy what was suitable to the existing art of medicine, for they discovered the epoch-making and fundamental notion of disease as a *natural process*, which can be identified, observed, explained as to its origin, predicted as to its outcome, and treated in ways conforming to this explanation, prediction, observation and identification.[100]

In the Coan treatises of the *Corpus*, diseases no longer occupy the foreground. Hippocrates' thought, in *Airs Waters Places*, is concerned first and foremost with human populations (and the sub-groups thereof) to be found in conditioning natural environments, and only *subsequently* with the concomitant diseases typical of the human types and sub-types that can be distinguished. Coan medicine, especially in its fine flowering in the early books of the *Epidemics*, which I do not hesitate to ascribe to Hippocrates,[101] is a medicine concerned with patients – both groups and individual persons – in a definitive environment, i.e., with sick human beings, not with diseases which, so to speak, would be independent and autonomous entities, as those classified in *Diseases* II A.

Thus, I would say, the traditional theorizing about the distinction between patient-oriented Hippocratic medicine on the one hand and disease-oriented Cnidian medicine on the other[102] has a *fundamentum in re*, howevermuch the degree of difference may have been exaggerated in modern times. It can hardly be denied, however, that the pendulum, which in the first half of the 19th cent. had moved in the direction of Hippocratic, i.e., patient-oriented, medicine, has in our own time swung back (or forwards?) to a disease-oriented medicine which is much more 'Cnidian' than 'Coan'. I have learnt much both from Professor Robert Joly and Professor Wesley D. Smith. However, any evaluation of the Hippocratic tradition, or of the scientific level of the early treatises of the *Corpus*, which not only refuses to acknowledge that an important distinction between Coans and Cnidians really exists, but also omits to take the development of medicine from the early 19th cent. up to our own times into account, is bound to remain too narrow.

Rijksuniversiteit Utrecht, Netherlands

NOTES

1 Hereafter J. Unless indicated otherwise, references by page numbers are to the present volume.

2 Lloyd, 1975, 176 f. n. 7. My italics. Deichgräber, 1933/1971a, 161 f., accepts that Ctesias mentioned Hippocrates.

3 See Wellmann, 1901, 51 ff., but compare Lloyd, 1975, 177 f.

4 Wellmann, 1901, 64.

5 Overrated by Wellmann anyhow; see Deichgräber, 1961, 34 f.

6 *Nutriment* (Wellmann, 1901, 52 f., but cf., e.g., Deichgräber, 1973), and *Sevens* (Wellmann, 1901, 55, but cf. Mansfeld, 1971).

7 Wellmann, 1901, 55 (Stephanus (?), *Schol. in Hipp. et Gal.*, II 326 Dietz: *τί φῄς, ὦ Ἱππόκρατες κτλ.*), and *ibid.*, 58 (Galen, XVII A 223 Kühn = *CMG* V 10, I 112, 31 f.: *ἐπὶ τίσι γὰρ ἐρεῖς κτλ.*). Discussed by Lloyd, 1975, 177 f. It should be noted, however, that of the works on Wellmann's list the two at issue here, viz., *Aph.* II and *Epid.* I, need not be denied to Hippocrates; for *Epid.* I, at any rate, see *supra*, p. 60, and n. 101. Yet we shall only be in a position to accept Galen's testimony if the authenticity of *Epid.* I can be established on other grounds. The testimony of Stephanus (?), who is fond of making up little dialogues, remains suspect.; Deichgräber, 1933/1971a, 160 n. 2, rejects the Stephanus(?)-passage, but tends to accept (*ibid.*, 160) that in Galen, without, however, concluding that Diocles said that *Epid.* I is by Hippocrates.

8 Soranus ed. Ilberg, *CMG* IV, Leipzig-Berlin, 1927, 175 f.

9 See now Grensemann, 1969, 71 f., and Langholf, 1977, 15 f. The pioneering study is that of Deichgräber, 1933/1971a, 16, 74 f., 144 f.

10 See *supra*, n. 9.

11 Ps. Soranus hardly cites his early sources at first hand, however; see *supra*, pp. 53–4.

12 For this custom see Mansfeld, 1980b, 86 f.

13 For lines 3–4 of the epigram, see *supra*, p. 66.

14 Otherwise unknown.

15 *CMG* IV, p. 175, 9 f. Ilberg: *κατὰ δὲ τοὺς Πελοποννησιακοὺς ἤκμασε χρόνους, γεννήθεις, ὡς φησὶν Ἰσχόμαχος ἐν τῷ πρώτῳ Περὶ τῆς Ἱπποκράτους αἱρέσεως, κατὰ τὸ πρῶτον ἔτος τῆς ὀγδοηκοστῆς Ὀλυμπιάδος, ὡς δὲ Σωρανὸς ὁ Κῷος, ἐρευνήσας τὰ ἐν Κῷ γραμματοφυλακεῖα, προστίθησι, μοναρχοῦντος Ἀβρίαδα, μηνὸς Ἀγριανίου ἑβδόμῃ καὶ εἰκοστῇ, παρ' ὃ καὶ ἐναγίζειν ἐν αὐτῇ μέχρι νῦν Ἱπποκράτει φησὶν τοὺς Κῴους.*

16 I have taken this term from Mosshammer's splendid study (1979), 159 f. and *passim.*

17 See Mosshammer, 1979, 158 f., and for Apollodorus' methods and reliability *ibid.*, esp. 113 ff., and Mansfeld, 1979.

18 The Athenian genealogist, first half fifth cent. BCE (passage from ps. Soranus at *FGrH* 3 F 59). Jacoby, in his *Commentary*, points out that Hippocrates is too late to have been included by Pherecydes; therefore, Eratosthenes (whom he supposes to have cited Pherecydes) could only have adduced the genealogist for the earlier part of the genealogy. Jacoby unnecessarily suggests that Pherecydes only gave the mythical part of the genealogy; but Pherecydes was a genealogical *historian*, who, if possible, brought down genealogies to his own day. This is important, because it shows that, already in the days of Hippocrates' father, the Asclepiads were prominent. This, again, confirms J.'s observation (p. 34) "Because a genealogy going back to the gods is legendary, this

does not make it legendary as well when it goes from grandfather, to father, to the person's children."

[19] Late first cent. CE. This gives us the *t.p.q.* of ps. Soranus.

[20] See Jacoby, 1902/1973, 29, 55, and the examples added Mansfeld, 1979, 58 f.

[21] Compare, for instance, Diogenes Laertius' *Life of Plato*. The genealogy is given at III 1, the chronology (for which Apollodorus is cited, *FGrH* 244 F 37) in the second half of III 2, without connection with the genealogy.

[22] See *FGrH* 244 F 14 (Crates and his pupils), F 30 (Parmenides-Zeno), F 34 (Anaxagoras-Socrates and -Euripides), F 36 (relative chronology Anaxagoras-Democritus), F 38 (Plato-Aristotle), F 41 (Nausiphanes-Epicurus), F 42 (Epicurus-Hermarchus), F 47, F 53–60 (various Academics).

[23] A testimony some years earlier than ps. Soranus, Celsus, *pr.* 8 (*CML* I, p. 18, 11 f. Marx) has: ... *et Democritus. Huius autem, ut q u i d a m crediderunt, discipulus Hippocrates*. The *Suda*, s.v. 564 Ἱπποκράτης, I 2, p. 662, 13 f. Adler has ὡς δέ τινες Δημοκρίτου τοῦ Ἀβδηρίτου· ἐπιβαλεῖν γὰρ αὐτὸν νεῷ πρεσβύτην. Thus, there is unanimity in the way Celsus, ps. Soranus, and the *Suda* present the heterodox opinion. De Ley 1969, 52, suggests that the heterodox view derives from a "niet-medische bron". This may be true, although Soranus (who lived about half a century later than Celsus) had no objection against enumerating Democritus among Hippocrates' pupils, see Tzetzes, *Chil.* IX, 951 f. and 988. I suggest, however, that the heterodox view, whatever its source, was preserved in the chronographical vulgate, which must have been known to Celsus and which, of course, is very important for the *Suda*. – For ἐπιβαλεῖν + indication of pupil-teacher relationship see Mansfeld, 1979, 42 n. 9; in the passages there quoted, the subject of the verb is always the pupil. If we apply this to the *Suda*-passage, Hippocrates, in his old age (!), would have become the pupil of young Democritus; perhaps, however, the sense is that Democritus' old age coincided with Hippocrates' youth, which would agree with chronographic parlance. As Kranz points out, ad *Vorsokr.* 68 A 10, Diels' emendation αὐτῷ νέον πρεσβύτῃ is no improvement.

[24] Cf. *FGrH* 244 F 32, F 34, F 47; for F 31, see Mansfeld, 1979, 58 f.

[25] Cf. *FGrH* 244 T 2, ps. Scymnus, *Orb. descr.* 30 ἐπιφανῶν ἀνδρῶν βίους.

[26] Edelstein, 1935, 1295 f., accepts Eratosthenes' and Apollodorus' "Angaben über Chronologie und Lebenszeit". This implies that he attributes Ischomachus' dates to the vulgate. Similarly Jacoby, 1902/1973, 295 ff. See next note.

[27] In his commentary on *FGrH* 244 F 73.

[28] Cf. Jacoby, 1902/1973, 297: "Pseudo-Soranus hat das von Istomachus [so the earlier editions of the *Life*, still followed by Jacoby also in the text of *FGrH* 244 F 73a] gegebene geburtsjahr angenommen und hat es genauer definiert [viz., by quoting Soranus of Cos], was uns freilich nicht viel hilft, da das jahr des Abriadas weder uns noch dem Soran bekannt war" [see however *infra*, n. 43]. Edelstein, 1935, 1293, is vague about the distinction between Ischomachus and Soranus of Cos, and wrong *ibid.*, 1296, 15 f.: "Das genaue Datum der Geburt, welches Ischomachos *und* Soran von Kos aus den koischen Archiven geben . . . " (my italics).

[29] See Jacoby, cited *supra*, n. 28 (he thinks Soranus may be "glaubwürdig"). A similar reserve in Edelstein, 1935, 1296.

[30] Sherwin-White, 1978, 189–91, discusses Soranus' evidence and rejects it as a 'worthless anachronism'. I cite Ms Sherwin-White's main arguments in the text.

[31] Cf. Sherwin-White, 1978, 191 f.

32 *O.c.*, 387.
33 Cf. Sherwin-White, 1978, 61.
34 P. 175, 14 f. Ilberg: παρ' ὃ καὶ ἐναγίζειν ἐν αὐτῇ μέχρι νῦν Ἱπποκράτει φησὶ τοὺς Κῴους.
35 Edelstein, 1935, 1299. But ἐναγίζειν refers to the sacrifices offered to dead persons generally, not necessarily to heroes (although the dead, in time, could be promoted to heroes). See Burkert, 1977, 299, 307, 316, 405.
36 See Ziebarth, 1910, 24, 160, and cf. Mansfeld, 1980b, 86. With Soranus' (*supra*, n. 34) μέχρι νῦν cf. Alcidamas *ap*. Arist., *Rhet*. B 23, 1398b 16 f., Λαμψακηνοὶ Ἀναξαγόραν ξένον ὄντα ἔθαψαν καὶ τιμῶσι ἔ τ ι κ α ὶ ν ῦ ν, and Diog. Laert. II 14 φ υ λ ά τ τ ε τ α ι τὸ ἔθος κ α ὶ ν ῦ ν.
37 Boyancé, 1936/1972, 324.
38 Philodemus, *De piet*., p. 94 (ἑορτῶν καὶ θυσιῶν etc.), p. 126, p. 127 (πάσαις ταῖς πατρίοις ἑορταῖς κε[χ]ρ[η]μένος). See Boyancé, 1936/1972, 325 and n. 2.
39 Still observed in the first cent. BCE (Cic., *Fin*. II 101) and the first cent. CE (Pliny, *NH* XXXV 5).
40 Burkert, 1977, 316, and Nilsson, 1961, 143 n. 8, point this out as regards heroes. For Anaxagoras see *supra*, n. 36. – Note that Hippocrates' tomb was in Thessaly, not Cos.
41 It may, or may not, be important that the keepers of legal documents concerned with private citizens are called *chreophylakes*; see Sherwin-White, 1978, 213.
42 See *supra*, n. 31 and text thereto, for arguments that the institution precedes the synoecism of 366 BCE.
43 Consult Mosshammer's splendid discussion, 1979, 94 f., and cf. his statement, *ibid*., 96: "The annual lists of magistrates, kings, priests, and victors on which early Greek chronology is based are structurally sound".
44 For Epicurus' birth-day, which never became an official state festival, see *supra*, p. 55. It is noteworthy that Apollodorus, *FGrH* 244 F 42, gives the year (Athenian archon!), month and day of Epicurus' birth [there is a slight discrepancy, as to the day, with Epicurus' will – which of course does not mention the year of birth –; see Jacoby, *Comm., ad loc.* Personally, I believe that the number-word in the transmitted text of Apollodorus is corrupt]; he *computed* the year, and will have known the month and day because he knew of the Epicurean celebration (Apollodorus was an Athenian himself). Apollodorus also gave the year, month and day of Socrates (F 34) and Plato (F 37); here, too, he *computed* the year, and must have known about the date from another source. Wilamowitz-Moellendorff, 1920, 272, plausibly suggests that Socrates' birthday (6th Thargelion) and Plato's (7th Thargelion) were celebrated in the Academy on successive days; indeed, the most probable source of information for Apollodorus is the observation of a custom about which he, an Athenian, must have known. For Plato's and Socrates' birthday see Boyancé, 1936/1972, 259 ff. I accept the argument of Lynch, 1972, 108 ff., against Wilamowitz and many others (among whom Boyancé), who held that the Academy was a religious association devoted to the Muses (a *thiasos*). Lynch, 1972, 118, is right in pointing out that Epicurus' school (*pace* Wilamowitz, also a *thiasos*) was not devoted to a cult of the Muses, but to a "cult . . . of Epicurus himself". But the fact that neither the Academy nor the Kepos are *thiasoi* does not preclude – as Lynch acknowledges himself in the case of Epicurus – a private 'cult', i.e., the celebration of birth-days in commemoration of persons very important to the celebrating society. On this point, some of Boyancé's arguments remain pertinent. Cf. also Müller, 1975, 20 and n. 1.

45 Published and commented upon by Bousquet, 1956.

46 ἔ δ ο ξ ε Ἀσκλαπιαδᾶν τῶι κοινῶι Κώιων καὶ Κνιδίων. A well-organized society!

47 Ἀσκλαπ[ιά]δας κατὰ ἀνδρο[γέν]ειαν. Cf. Bousquet, 1956, 587.

48 For their genealogies as known to prominent Coans in the first cent. CE, see Sherwin-White 1978, 258.

49 Plato, *Prot*, 311 f. Note that the Hippocratic *Oath* obliges the person who pronounces it to the following: "that I shall hold my teacher in this art equal to my own parents; that I shall make him partner in my livelihood; that, when he is in need of money, I shall share mine with him; that I shall consider his offspring (γένος) as my own brothers, and shall teach them this art, if they express a desire to learn it, without fee or indenture; that I shall impart precept, oral instruction and everything else that is to be learnt to my sons, to the sons of my teacher, and to indentured pupils who have taken the physician's oath, but to nobody else" (tr. Jones, slightly modified). It is interesting to see that this oath is destined for outsiders only; apparently, these must declare themselves bound to all the obligations which are accepted by true-blue Asclepiads, without, however, achieving completely similar status, for they have to teach their own sons, which, accordingly, can be refused by members by descent (and they cannot become surgeons either). The physician who is a former indentured pupil shall never be an Ἀσκλαπιάδης κατ' ἀνδρογένεαν. For the *koinon* and its later history see further Sherwin-White, 1978, 257 ff.; for the section of the oath I have cited see Deichgräber, 1933/1971b, 100 f.

50 *mox adventu Aesculapii artem medendi inlatam maximeque inter posteros eius celebrem fuisse, nomina singulorum referens quibus quisque aetatibus viguissent* (*viguissent* = ἤκμασαν, and presumably refers to their *floruit*).

51 Sherwin-White, 1978, 283 f.

52 See Bardong, 1942, 577 ff.; Langholf, 1977, 19.

53 Diller, 1959/1971, 40 f. argues from II–IV–VI and Larissa to I and III (although his argument is not very clear).

54 See Bardong, 1942, *passim*; Diller, 1959/1971, 41.

55 Cf. the evidence referred to Mansfeld, 1971, 25 f. n. 116, and Harig, 1980, esp. 239: "relativ spätes Werk aus der ersten Hälfte des 4. Jahrh."

56 Joly, 1960, esp. 203–209; cf. also the introduction to his edition of this work in the Budé series.

57 Smith, 1979, 48.

58 I have tried to make up for this omission in a recently published paper (Mansfeld, 1980c), the substance of which was part of my communication at Montréal.

59 See now Mansfeld, 1980c, 344 ff.

59a Since this was written, Joly kindy sent me an advance copy of a manuscript (Aug. 1981) in which he attempts a refutation; I shall reply to this in due time.

60 See *supra*, p. 51.

61 See *supra*, n. 52 and text thereto; for the date, see *supra*, p. 51.

62 For such differences see e.g. Deichgräber, 1933/1971a, 126. Apart from the author's development, also such differences as are a consequence of the distinct *genres* to which these works belong should be taken into account.

63 Diller, 1959/1971, 39 ff.

64 Diog. Laert. III 37. On the history of the *Corpus Platonicum* see Müller, 1975, 22ff. For a comparison between Aristotelian and Hippocratic treatises see Wilamowitz, [3]1912, 100.

[65] Note that Protagoras, according to Plato, attacked, in writing, "all the arts": *Sph.* 232 a–e = *Vorsokr.* 80 B8. Cf. also Heinimann, 1976, 127 ff. For the variety of genres in the *Corpus Hippocraticum* see Wilamowitz, ^{3}1912, 99–100.

[66] Cf. also J., p. 32. In *Theaetetus*, Protagoras' treatise has been read by both Socrates and young Theaetetus (*Tht.* 152 a); yet it was, apparently, less well-known than *Airs Waters Places*, for Socrates has Theaetetus confirm explicitly that he has read it. That Phaedrus, in *Phaedrus*, is himself a person with medical connections should not be interpreted as implying that this is why he knew Hippocrates' work, for (1) Socrates knew it, too, and (2) Plato had to assume that his allusions to it would present no difficulty to the contemporary reader. [I intended to deal with medicine and rhetoric in *Phaedrus* elsewhere].

[67] Cf. e.g., Langholf, 1977, and Bardong, 1942.

[68] See Düring, 1968, 259 f., 313 f., and Peck, 1965, LIV f. The exceptional case of the *Problemata* is even more glaring: Aristotle's own *Probl.* are lost, and the work we have is a Peripatetic product of the mid-third cent. BCE, which only contains possible echoes of Aristotle's; the ancients, however, also knew genuine 'Aristotelian' *Problemata*. (Flashar, 1962, 303 ff.) – A parallel from the sphere of works written for publication is the *Epinomis* which Philip of Opus added to the *Laws* he edited after Plato's death; Müller, 1975, 25, points out that the criteria for addition or inclusion applied by members of a school are not philological, but bound up with considerations of what is useful.

[69] *Epid.* V and VII (parts of which coincide) are to be dated to c. 360–336 BCE (Langholf 1977, 16 f.).

[70] Grensemann, 1969, 77 ff.

[71] This observation should discourage those who want to ascribe a purely empirical methodology, so-called, to the author(s) of *Epid.* I and III; see Grensemann, 1969, 80, whose interpretation of *τὰ γεγραμμένα*, however, I do not follow.

[72] Deichgräber's research (1933/1971a) was independent of Galen's information, but is confirmed by it, see Bardong, 1942. Cf. also Langholf, 1977, 16.

[73] Diller, 1959/1971, 39, who points out that this digest, with other similar collections of abstracts, proves the existence of a body of works which 'in einem bestimmten Zusammenhang miteinander standen". Ibid., n. 14, he refers to O. Poeppel's dissertation (*non vidi*).

[74] See Deichgräber, 1933/1971a, 171: *Diseases* I–II–III. Among the Coan works abstracted are *Epid.* II, VI, VII and *Prognosticon*; there are also abstracts from *Regimen in Acute Diseases*, whose status as a Coan treatise is unclear.

[75] Diller, 1959/1971, 43.

[76] *Supra*, pp. 56–7.

[77] Diller, 1959/1971, 30.

[78] See previous note.

[79] Müller, 1975, 23 f.

[80] *Supra*, pp. 50–1.

[81] *Supra*, p. 50.

[82] See Düring, 1968, 313 f.

[83] Cf. also p. 41, on the 'critical days', where J. successfully appeals to Bachelard's famous *obstacles épistémologiques*, i.e., the *unconscious* factors which obstruct the progress of science (note the subtitle of Bachelard, 1938: "Contribution à la psychanalyse de la connaissance objective"), Bachelard explicitly refused to consider such obstacles

as he calls "external", and similarly refused to blame either our intellect or our senses for their shortcomings. To my knowledge, however, it has not been noticed that *obstacles épistémologiques* can be translated into Greek: *τὰ κωλύοντα εἰδέναι*, words which occur in a seminal fragment of the fifth-cent. BCE Sophist Protagoras, *Vorsokr.* 80 B4. Here it is said that such obstacles are "many", the "obscurity of the subject" and the "shortness of human life" being given as instances. Protagoras' thought was important to the medical writers in many ways; here, it is sufficient to point at the awareness of epistemic obstacles as visible in the first *Aphorism* (where they are not as total as in Protag., *loc. cit.*): the first difficulty listed in the aphorism is that "life is short" (*vita brevis*). On Protagoras see further Mansfeld, 1981.

[84] Smith, 1973, 569 ff.

[85] Joly, 1966, 64 ff.

[86] *Ibid.*, 69. My italics. 'Mentalité scientifique', in Joly, 1966, means *prescientific mentality.*

[87] 421.

[88] 422.

[89] *Supra*, pp. 59–60, and Mansfeld, 1980c, 352 ff. Note that J., in his present paper, still holds the view propounded in his 1961 paper, viz., that Plato, in *Phaedrus*, means environmental medicine.

[90] *Acut.* 1–3 = T10 Grensemann (Grensemann, 1975). See also next note.

[91] For the fragment of Euryphon (= T15 Grensemann) see also Jouanna, 1974, 17 f. On *Morb.* II A in general see Jouanna, *ibid.*, 26 ff., 163 ff. On *Acut.* 1–3 see also Kudlien, 1977, 94 f. Di Benedetto, 1980, 109, accepts both the testimony of *Acut.* 1–3 and, *ibid.*, 105, Jouanna's argument concerned with *Morb.* II A, without noticing that this undermines his general argument that a distinction between schools of Cos and Cnidus is impracticable.

[92] This distinction is also consistent with the preference for *νόσημα* ("diseased condition"), not *νοῦσος* ('disease'), which, according to G. Preiser, 1976, is a distinctive characteristic of Coan works as opposed to Cnidian ones.

[93] Sc., of Presocratic nature; see Mansfeld, 1980c, 354 f.

[94] This is postulated by di Benedetto, 1980, too.

[95] I do not object to the assumption that he is; see my paper 1980a, 385.

[96] For criticism of this notion see *supra*, n. 83.

[97] *οὐ τυχῇ, ἀλλα τεχνῇ.* For the epigram see *supra*, p. 52.

[98] Cf. Mansfeld, 1980c, 354 ff., 360 f.

[99] Cf. Mansfeld, 1980c, 359 f.

[100] See Mansfeld, 1980b, 378 ff.

[101] *Supra*, p. 60. Robert, 1978, although not committing himself as to questions of authorship, beautifully shows that *Epid.* I–VI belongs with *Airs Waters Places*, even where the details are concerned.

[102] See Lonie, 1978, and Smith, 1979, Chapter I, *passim*, who impressively study this history (although additions could be made, esp. as to the history of Hippocratism in Germanic countries).

BIBLIOGRAPHY

Bachelard, G.: 1938, *La formation de l'esprit scientifique*, Vrin, Paris.

Bardong, K.: 1942, 'Beiträge zur Hippokrates- und Galenforschung, I. Das kleine Notiztafelchen des Hippokrates'. *Nachrichten Akademie Göttingen*, 577–603.

Benedetto, V. di: 1980, 'Cos e Cnido', in: Grmek 1980, 97–112.

Bourgey, L. and Jouanna, J. (eds.): 1975, *La collection hippocratique et son rôle dans l'histoire de la médecine*, Colloque de Strasbourg (23–27 octobre 1972), Leiden, Brill 1975.

Bousquet, J.: 1956, 'Delphes et les Asclépiades', *Bulletin de Correspondance Hellénique* 80, 579–593.

Boyancé, P.: 1936/1972, *Le culte des Muses chez les philosophes grecs*, De Bocard, Paris.

Burkert, W.: 1977, *Griechische Religion der archaischen und klassischen Epoche*, Kohlhammer, Stuttgart etc.

Classen, C. J. (ed.): 1976, *Sophistik*, Wege der Forschung Bd. 187, Wissenschaftliche Buchgesellschaft, Darmstadt.

Deichgräber, K.: 1933/1971a, *Die Epidemien und das Corpus Hippocraticum*, De Gruyter, Berlin.

Deichgräber, K.: 1933/1971b, 'Die ärztliche Standesethik des hippokratischen Eides', in: Flashar, 1971, 94–120.

Deichgräber, K.: 1961, 'Vindicianus', in: *Pauly-Wissowa*, 2. Reihe, 17. Halbband, 29–36.

Deichgräber, K.: 1973, *Pseudhippokrates über die Nahrung. Eine stoisch-heraklitisierende Schrift aus der Zeit um Christi Geburt*, Abhandlungen Akademie Mainz, Steiner, Wiesbaden.

Diller, H.: 1959/1971, 'Stand und Aufgaben der Hippokratesforschung', in: Flashar 1971, 29–51.

Düring, I.: 1968, 'Aristoteles', in: *Pauly-Wissowa*, Supp. Bd. XI, 159–336.

Edelstein, L.: 1935, 'Hippokrates', in: *Pauly-Wissowa*, Supp. Bd. VI, 1290–1345.

Flashar, H.: 1962, *Aristoteles, Problemata physica-Aristoteles. Werke in deutscher Übers.* hrsg. v. E. Grumach, Bd. 19, Wissenschaftliche Buchgesellschaft, Darmstadt.

Flashar, H.: 1971, *Antike Medizin*, Wege der Forschung Bd. 221, Wissenschaftliche Buchgesellschaft, Darmstadt.

Grensemann, H.: 1969, 'Die Krankheit der Tochter des Theodoros. Eine Studie zum siebten hippokratischen Epidemienbuch', *Clio Medica* 4, 71–83.

Grensemann, H.: 1975, *Knidische Medizin*, I, *Die Testimonien zur ältesten knidischen Lehre und Analysen knidischer Schriften im Corpus Hippocraticum*, Ars Medica II, Bd. 4, 1, De Gruyter, Berlin, New York.

Grmek, M. D. (ed.): 1980, *Hippocratica*, Actes du Colloque hippocratique de Paris (4–9 septembre 1978), Editions du CRNS, Paris.

Harig, G.: 1980, 'Anfänge der theoretischen Pharmakologie im Corpus Hippocraticum', in: Grmek, 1980, 233–246.

Heinimann, F.: 1961/1976, 'Eine vorplatonische Theorie der Techne', in: Classen, 1976, p. 127–169.

Hinneberg, P. (ed.): ³1912, *Die Kultur der Gegenwart*, Teil I Abt. VII. Die griechische und lateinische Literatur und Sprache, Teubner, Leipzig-Berlin.

Joly, R.: 1960, *Recherches sur le traité pseudo-hippocratique Du Régime*, Les Belles Lettres, Paris.

Joly, R.: 1961, 'La question hippocratique et le témoignage du Phèdre', *Revue des Etudes Grecques* 74, 69–72.

Joly, R.: 1966, *Le niveau de la science hippocratique. Contribution à la psychologie de l'histoire des sciences.* Les Belles Lettres, Paris.

Joly, R.: 1971, 'Die hippokratische Frage und das Zeugnis des Phaidros', in: Flashar, 1971, 52–82 (= Joly, 1961).

Joly, R.: 1972, 'Hippocrates', in: *Dictionary of Scientific Biography*, Vol. VI, New York, 418–431.

Joly, R. (ed.): 1977, *Corpus Hippocraticum*, Colloque de Mons (septembre, 1975), Editions Universitaires de Mons, Mons.

Joly R.: 1980, 'Un peu d'épistémologie historique pour hippocratisants', in: Grmek 1980, 285–298.

Jacoby, F.: 1902/1973, *Apollodors Chronik. Eine Sammlung der Fragmente*, Weidmann, Berlin / Arno Press, New York.

Jouanna, J.: 1974, *Hippocrate. Pour une Archéologie de l'Ecole de Cnide*, Les Belles Lettres, Paris.

Kerferd, G. B. (ed.): 1981, *The Sophists and their Legacy*, Proceedings of the fourth international colloqiuum on Ancient Philosophy, Bad Homburg 19th Aug. – 1st Sept. 1979, Steiner, Wiesbaden.

Kudlien, F.: 1977, 'Bemerkungen zu W. D. Smith's These über die knidische Ärzteschule', in: Joly, 1977, 95–103.

Langholf, V.: 1977, *Syntaktische Untersuchungen zu Hippokrates-Texten*, Abhandlungen Akademie Mainz, Steiner, Wiesbaden.

Lloyd, G. E. R.: 1975, 'The Hippocratic Question', *Classical Quarterly* N. S. **25**, 171–192.

Lonie, I. M.: 1978, 'Cos *vs* Cnidus and the Historians', *History of Science* **15**, 42–75, 77–92.

Lynch, J.: 1972, *Aristotle's School. A Study of a Greek Educational Institution*, University of California Press, Berkeley etc.

Mansfeld, J.: 1971, *The Pseudo-Hippocratic Tract* ΠΕΡΙ 'ΕΒΔΟΜΑΔΩΝ *Chapters 1–11 and Greek Philosophy*, Van Gorcum, Assen.

Mansfeld, J.: 1979, 'The Chronology of Anaxagoras' Athenian Period and the Date of his Trial', Pt. I, *Mnemosyne* Ser. IV **32**, 39–69.

Mansfeld, J.: 1980a, 'The . . . Trial', Pt. II, *Mnemosyne* Ser. IV **33**, 17–95.

Mansfeld, J. 1980b, 'Theoretical and Empricial Attitudes in Early Greek Scientific Medicine' in: Grmek, 1980, 371–392.

Mansfeld, J.: 1980c, 'Plato and the Method of Hippocrates', *Greek, Roman and Byzantine Studies* **21**, 341–362.

Mansfeld, J.: 1981, 'Protagoras on Epistemological Obstacles and Persons', in: Kerferd, 1981, 38–53.

Mosshammer, A. A.: 1979, *The Chronicle of Eusebius and Greek Chronographical Tradition*, Bucknell University Press, Lewisburg-London.

Müller, C. W.: 1975, *Die Kurzdialoge der Appendix Platonica. Philologische Beiträge zur nachplatonischen Sokratik*, Fink, München.

Nilsson, M. P.: ²1961, *Geschichte der griechischen Religion*, II, Beck, München.

Pauly-Wissowa: 1894–1978, *Pauly's Realencyclopädie der classischen Altertumswissenschaft*. Neue Bearbeitung begonnen von Georg Wissowa, fortgeführt von W. Kroll und K. Mittelhaus, hrsg. von K. Ziegler, Metzler, Stuttgart-Druckenmüller, München.

Peck, H. A.: 1965, *Aristotle. Historia Animalium.* Vol. I, Loeb Classical Library, Heinemann, London.

Preiser, G.: 1976, *Allgemeine Krankheitsbezeichnungen im Corpus Hippocraticum. Gebrauch und Bedeutung von nousos und nosema*, Ars Medica II, Bd. 5, De Gruyter, Berlin.

Robert, F.: 1975, 'Les adresses des malades dans les Epidémies II, IV et VI', in: Bourgey-Jouanna 1975, 173–194.

Sherwin-White, S. M.: 1978, *Ancient Cos. An Historical Study from the Dorian Settlement to the Imperial Period*, Vandenhoeck & Ruprecht, Göttingen.

Smith, W. D.: 1973, 'Galen on Coans vs Cnidians', *Bulletin History of Medicine* **47**, 569–585.

Smith, W. D.: 1979, *The Hippocratic Tradition*, Cornell University Press, Ithaca and London.

Wellmann, M.: 1901, *Die Fragmente der griechischen Ärzte*, I. *Die Fragmente der sikelischen Ärzte Akron, Philistion, und des Diokles von Karystos*, Weidmann, Berlin.

Wilamowitz-Moellendorff, U. von: [3]1912, *Die griechische Literatur und Sprache*, in: Hinneberg [3]1912, 3–318.

Wilamowitz-Moellendroff, U. von: 1920, *Platon*, I, Weidmann, Berlin.

Ziebarth, E.: [2]1910, *Aus dem griechischen Schulwesen*, Teubner, Leipzig-Berlin.

PART II

JOHN BEATTY

WHAT'S IN A WORD? COMING TO TERMS IN THE DARWINIAN REVOLUTION

INTRODUCTION

For all its rigor, Darwin's *Origin* is constructed in a very peculiar way. It is, then again it is not, clear what the *Origin* is about. Of course, it is apparently about the origin and evolution of species. On the other hand, early on and throughout the *Origin*, Darwin denies the reality of species. We are thus confronted with the perplexing proposition that species originate and evolve naturally, though species are not real.

In order to understand how this situation came about, and how it was rectified by Darwin's successors, it is necessary to take into account certain respects in which theory change is affected by the theory-laden meanings of scientific terms. In a nutshell, given the loaded meaning of 'species' derived from nonevolutionary theories of natural history, Darwin had good reason to question the reality of species – just as Lavoisier had good reason to question the reality of phlogiston, given the theory-laden definitions of 'phlogiston'. But the term 'species', unlike the term 'phlogiston', was not elminated from the scientific vocabulary when it was discovered that nothing existed corresponding to the traditional definition. Having chosen to discuss the evolution of species by natural selection, it was then up to Darwin's successors to figure out what in the world his theory was about. They accomplished this by redefining 'species' in terms of the new theory, and by thus explaining the new sense in which species are real.

By way of historiographic introduction, I am well aware of the contempt of historians for philosophers who, in turn, see the history of science as a stockpile of case studies for current views of theory change. There is some concern that current views of science are being "imposed harshly and anachronistically on the scientific events of the past," as I. B. Cohen has said specifically of Kuhnian analyses of the history of science (Cohen, 1976, p. 53). The philosophy of science relevant to an event in the history of science, historians often argue, is the philosophy of science contemporaneous with that event. So the only philosophy of use to most historians is history of philosophy – the most current philosophy being appropriate for understanding only the most current historical events. On what grounds, then, can

Michael Ruse (ed.), Nature Animated, 79–100.

contemporary-sounding considerations, like the effect of theory-laden meaning on theory change, be legitimately brought to bear upon the understanding of past developments, like the history of evolutionary theory?

In the first place, there is no reason why contemporary philosophers cannot discover aspects of theory change which were operative in the past, though previously overlooked by philosophers and scientists. Thus, for instance, the fact that nineteenth-century British notions of a scientific revolution differ significantly from Kuhn's notion, does not itself preclude the occurrence of a Kuhnian revolution in nineteenth-century British science. Similarly, if nineteenth-century scientists and philosophers of science did not discuss the theory-laden meaning of scientific terms, it may simply be that they overlooked an important, operative aspect of theory change in the nineteenth century.

On the other hand, *had* the effects of theory-laden meaning on theory change actually been at issue in nineteenth-century philosophy and science, a historical analysis of the period in those terms would be less likely to be anachronistic. For in that case, the analysis would be less likely to attribute to the science of the past, characteristics derived from, and more properly attributed to, contemporary science. In that case, too, the analysis would be less likely to attribute to scientists of the past, rationales which they themselves could not or would not have recognized. As it happens, theory ladenness *was* treated in depth in the philosophy of science of the period. Moreover, many evolutionists and nonevolutionists were aware that nonevolutionary definitions of 'species' placed significant constraints on the formulation of evolutionary theory.

Taking into account not only the constraints of theory-laden language on theory change, but also the recognition of those constraints by nineteenth-century philosophers and scientists, this analysis of the history of evolutionary theory bears some similarity in strategy to Maurice Crosland's *Historical Studies in the Language of Chemistry*. As Crosland prefaced his study,

> ... many of the old [chemical] names tended to perpetuate the misconceptions of a previous age about the nature of particular substances. It was difficult for each new generation to think afresh about the basic problem in chemistry, that of chemical composition, without carrying the prejudices implied in the current terminology. [But] ... the importance of language in the history of chemistry is not merely a twentieth-century idea conceived under the influence of the philosophical school of linguistic analysis; it has always been insisted upon by chemists themselves, men like Robert Boyle in the seventeenth century, Torbern Bergman and Lavoisier in the eighteenth century and Berzelius in the nineteenth century. (Crosland, 1962, pp. xiii-xiv)

THE PROBLEM

In the final pages of the *Origin*, Darwin concludes,

> When the views entertained in this volume on the origin of species, or when analogous views are generally admitted, we can dimly foresee that there will be a considerable revolution in natural history. Systematists will be able to pursue their labors as at present; but they will not be incessantly haunted by the shadowy doubt whether this or that form be in essence a species. This I feel sure, and I speak after experience, will be no slight relief. (Darwin, 1959, p. 484)

The reason that systematists would no longer bicker about whether fifty species of British brambles were really fifty, or really more or less, is that species would be considered arbitrary collections. There are simply no real species to bicker about. As Darwin continued, reemphasizing a position held throughout the *Origin,*

> In short, we shall have to treat species in the same manner as those naturalists treat genera, who admit that genera are merely artificial combinations made for convenience. This may not be a cheering prospect; but we shall at last be freed from the vain search for the undiscovered and undiscoverable essence of the term species. (Darwin, 1859, p. 485)

Darwin's position on the artificiality – nonreality – of species is perhaps not unreasonable in and of itself. But in conjunction with his theory of the evolution of species by natural selection, his position on species seems rather self-defeating. Who cares if species evolve by natural selection, if there's no such thing as a species?[1] It should be noted that this apparent confusion within Darwinian evolutionary theory is not a historical pseudoproblem. In the third volume of his *Contributions to the Natural History of the United States*, published just after the *Origin*, Louis Agassiz expressed precisely this sort of reservation concerning Darwin's theory:

> It seems to me that there is much confusion of ideas in the general statement, of the variability of species, so often repeated of late. If species do not exist at all, as the supporters of the transmutation theory maintain, how can they vary? And if individuals alone may exist, how can the differences which may be observed between them prove the variability of species? (Agassiz, 1860, pp. 89–90, n. 1)

A LINGUISTIC ANALYSIS OF THE PROBLEM

How did this situation come about? In large part, it was the result of language constraints on theory change: the terms available to Darwin and other evolutionists were loaded in their opponents' behalf. In the eighteenth and early nineteenth centuries, species were, by definition of the term 'species', constant in character; hence their reality implied their immutability.[2] These conditions are clearly not conducive to discussion of the evolution of species. For instance, how was one to argue for the evolution of species given Buffon's definition, according to which,

> We should regard two animals as belonging to the same species if, by means of copulation, they can perpetuate themselves and preserve the likeness of the species; and we should regard them as belonging to different species if they are incapable of producing progeny by the same means. (Buffon, 1749, p. 10; in Lovejoy 1959, p. 93)

On this definition, the continued existence of a species necessitated the preservation of its likeness. On this definition, then, a species simply could not evolve while continuing to exist.[3]

The nonevolutionary connotations of the term 'species' were drawn quite explicitly by Charles Lyell, who devoted considerable attention to the history of the organic world in his celebrated *Principles of Geology*. Insisting that it was necessary to consider the meaning of the term 'species' before considering whether species are modifiable, Lyell proceeded to convey the traditional meaning, loaded with the nonevolutionary assumptions of the time:

> The name of species . . . has been usually applied to "every collection of similar individuals produced by other individuals like themselves." This definition . . . is correct; because every living individual bears a very close resemblance to those from which it springs. But this is not all which is usually implied by the term species; for the majority of naturalists agree with Linnaeus in supposing that all the individuals propagated from one stock have certain distinguishing characters in common, which will never vary, and which have remained the same since the creation of each species. (Lyell, 1835, II, p. 407)

As if this definition of 'species' alone does not make it difficult enough to talk about the evolution of real species, Lyell also made explicit the connection between the reality of species and their modifiability. The question at issue, as he put it, is,

> . . . whether species have a real and permanent existence in nature? or whether they are capable, as some naturalists pretend, of being indefinitely modified in the course of a long series of generations? (Lyell, 1835, II, p. 405)

The choice offered by Lyell – between the reality of species and their modifiability – prefaced his discussion of Lamarck's theory of evolution, along with Lamarck's denial of the reality of species.[4] Lyell's treatment of

Lamarck, which Darwin read on the *Beagle*, served as Darwin's introduction to Lamarck's evolutionary theory. Later, when formulating his own evolutionary theory, Darwin was faced with the same choice – between the evolution of species and their reality.[5]

That the dichotomy between the reality and mutability of species actually constrained evolutionary thinking, has been pointed out on numerous occasions by Ernst Mayr. In one place, Mayr refers to this failure to distinguish reality from constancy as "one of the minor tragedies in the history of biology" (Mayr, 1957, p. 2), and in another place as a "violation of scientific logic" (Mayr, 1972, p. 987). But these epithets obscure the intrinsic place of such language constraints in theory change.

To the extent that scientific terms are theory laden – that is, to the extent that terms derive their meanings from the theories in which they are employed – it is, as Mayr is aware, more difficult to formulate alternative theories in those same terms. That scientific terms *are* theory laden has long been recognized. And no one has been more aware of theory ladenness than the influential Victorian historian and philosopher of science, William Whewell. It is not surprising that a historian and philosopher who coined so many of the scientific terms used by his contemporaries would be interested in the general manner in which scientific terms acquire their meaning.[6]

Whewell's reflections on the language of science convinced him that,

> . . . opinions, even of a recondite and complex kind, are often involved in the derivation of words; and thus . . . scientific terms, framed by the cultivators of science, may involve received hypotheses and theories. (Whewell, 1847, II, p. 491)

For example, Whewell noted, the term 'force' derives its precise meaning from Newton's first law of motion (Whewell, 1847, II, p. 488). That is, the definition of 'force' as "any cause which has motion or change of motion for its effect" is implicit in the law that "When a body moves not acted upon by any force, it will go on perpetually in a straight line and with a uniform velocity" (Whewell, 1847, I, pp. 216–217).

Thus, though Kepler and Newton both used the same term, 'force', they attached to it very different meanings (Whewell, 1837, II, p. 19). For Kepler assumed, contrary to the first law of motion, that force was required to maintain as well as to change motion. Kepler, then, as opposed to Newton, would have measured force by the velocity that a body *has*, rather than by the veolocity that a body *gains*. Interestingly, Whewell commented, with regard to the Keplerian use of 'force', "Such a use of language would prevent our obtaining any laws of motion at all" (Whewell, 1847, I, p. 266).

Whewell did not draw explicit connections between traditional definitions of 'species' and nonevolutionary theories of natural history; though the connections are at least as apparent as in the examples of theory ladenness he provided. The connections are, at any rate, implicit in various of his discussions of species. For instance, following Lyell, Whewell identified the reality of species with their immutability:

> . . . there is a capacity in all species to accommodate themselves, to a certain extent, to a change of external circumstances; this extent varying greatly according to the species. There may thus arise changes of appearance or structure, and some of these changes are transmissable to the offspring: but the mutations thus super-induced are governed by constant laws and confined within certain limits. Indefinite divergence from the original type is not possible; and the extreme limit of possible variation is reached in a short period of time: in short, *species have a real existence in nature,* and a transmutation from one to another does not exist. (Whewell, 1837, III, pp. 575–576; Whewell's emphasis)

Had Whewell explicitly considered the connections between traditional definitions of 'species' and nonevolutionary theories of natural history, he might also have added a comment similar to his remark about the Keplerian use of 'force': that such a use of language prevents our obtaining any laws of the evolution of species at all.

An opinion certainly stands a good chance of being preserved when the definitions of the terms of the opinion reflect the opinion itself, and as long as the terms of the opinion continue to be suitable terms of discourse. Under these conditions, Whewell noted,

> . . . the influence of preceding discoveries upon subsequent ones, of the past upon the present, is most penetrating and universal, though most subtle and difficult to trace. The most familiar words and phrases are connected by imperceptible ties with the reasonings and discoveries of former men and distant times. (Whewell, 1847, I, p. 271)

But is this influence of past science on present – this tendency to preserve the past in the present – a virtuous or a pernicious aspect of the theory ladenness of scientific terms? One's answer here depends on one's view of scientific progress. Has progress consisted mainly in the preservation of scientific opinions, by inductions from those opinions to present opinions? Or has progress consisted mainly in the replacement of older opinions by alternative opinions which are in some measure better? Whewell's view on this matter seems straightforward enough:

> Our examination of the history of science has led us to a view very different from that which represents it as consisting in the succession of hostile opinions. It is, on the

contrary, a progress, in which each step is recognized and employed in the succeeding one. Every theory, so far as it is true, (and all that have prevailed extensively and long, contain a large portion of the truth) is taken up into the theory which succeeds and seems to expel it. All the narrower inductions of the first are included in the more comprehensive generalizations of the second. And this is performed mainly by means of such terms as we are now considering – terms involving the previous theory. It is by means of such terms, that the truths at first ascertained become so familiar and manageable, that they can be employed as elementary facts in the formation of higher inductions. (Whewell, 1847, II, pp. 525–526)

So ties between present terminology and past opinion are welcome when present opinions are inductions from, or at least not alternatives to, past opinions. On the other hand, ties between present terminology and past opinion are unwelcome by investigators seeking serious alternatives to past opinion. Given the manner in which theory ladenness serves to preserve past opinion, how are proponents of serious alternatives to past opinion to proceed?

Proponents of alternative opinions may choose to replace the terminology of the past with new terminology that has no ties to the past. For instance, Whewell acknowledged that the term 'phlogiston', and its derivatives 'phlogisticated' and 'dephlogisticated', served to express a chemical theory that was rejected by succeeding generations of chemists (Whewell, 1847, II, p. 493). And he saluted the "courageous" and "foresighted" supporters of the oxygen theory for recoining the terms of chemistry (Whewell, 1947, II, p. 499). In response to Humphrey Davy's objection that "*oxygenated muriatic acid* is as improper a term as dephlogisticated marine acid," since 'a theoretical terminology is subject to continual alteration," Whewell argued that the terms in question, if improper, were not so because they involved theory, but because they involved false theory. And he added that the oxygen theory was so well established as to be considered a fact: "Is it not a fact that a combination of oxygen and hydrogen produces water?" (Whewell, 1847, II, p. 522–525).

Proponents of alternative opinions may also choose to retain, but redefine past terminology, in such a way that they can express alternatives to past opinions without contradicting themselves or without introducing conceptual confusions. Whewell at least implicitly acknowledged this means of severing the ties between present terminology and past opinion in his discussion of the differences between the Keplerian and Newtonian senses of 'force'. The laws of motion, he argued, could not have been expressed in terms of the Keplerian sense of 'force'.

It is the latter means of severing ties between the present and the past that must have been operative at some point in the Darwinian Revolution. For the term 'species', unlike the term 'phlogiston', was not eliminated from the scientific vocabulary, even though, as in the case of the elimination of 'phlogiston', it was discovered that nothing existed corresponding to the traditional definition of the term.

Interestingly enough, though, at least one evolutionist urged the elimination of the term 'species' on account of its nonevolutionary connotations. E. Ray Lankester is reported to have suggested discarding the term, on the grounds that "Modern zoology having abandoned Linnaeus' conception of species should ... abandon the use of the word." In Lankester's opinion, "the 'origin' of species was really the abolition of species" (Poulton, 1903, p. xci).

In any event, as Lankester seems to have recognized, Darwin had denied the reality of species, in the traditional sense of the term 'species', just by asserting their mutability. But Darwin himself did not redefine 'species' in such a way as to account for the reality of mutable species. Thus arose the conceptual difficulty concerning the evolution of unreal species. In fact, Darwin continued to employ the traditional definition of 'species', continuing to identify the reality of species with their immutability: "The power of remaining for a good, long period constant I look at as the essence of a species" (to Joseph Hooker, October 22, 1864, in F. Darwin, 1903, I, p. 252). Even in his reply to Agassiz's charge of conceptual confusion, Darwin relied upon the traditional definition of 'species':

> I am surprised that Agassiz did not succeed in writing something better. How absurd that logical quibble – 'if species do not exist, how can they very?' As if anyone doubted their temporary existence. (to Asa Gray, August 11, 1860, in F. Darwin, 1887, II, p. 124)

This rather fuzzy reply makes more sense when Darwin's reference to the temporary existence of species is interpreted as his acknowledgement that species are at least *temporarily constant*. But this interpretation of the reality of species does not leave much room for the reality of *evolving* species.[7]

That the old definition of species was unsuitable, in light of the new evolutionary theory, was recognized by Darwinians, if not by Darwin himself. Ernst Haeckel complained that the old definition was theoretically unsatisfactory, and that it led to circular reasoning about the immutability of species (Haeckel, 1879, pp. 50–51).[8] And after Lyell accepted Darwinian evolution (with the provision that it did not apply to man), he no longer identified the reality of species with their constancy. In the 1835 edition of *Principles of Geology*, recall, Lyell had asked,

. . . whether species have a real and permanent existence in nature? or whether they are capable . . . of being indefinitely modified in the course of a long series of generations? (Lyell, 1835, II, p. 405)

Whereas, in the 1872 edition, he asked,

. . . whether each species has remained from its origin the same, only varying within certain fixed and defined limits, or whether a species may be indefinitely in the course of a long series of generations. (Lyell, 1872, II, p. 247)[9]

The British naturalist, Henry Seebohm, went one step beyond recognizing that the old definition of 'species' was defunct. As he expressed the situation, simply, "The old definition of species having lapsed, in consequence of the rejection of the theory of special creation, it is necessary to provide one" (Seebohm, 1883, p. xi).

Where did the new definition come from? Well, to the extent that terms derive their meanings from the theories in which they are employed, the term 'species' ought to have derived its new meaning from the new evolutionary theory. That is, we would expect the new defining properties to reflect the properties that species must have in order that they might evolve by natural selection. A number of Darwin's contemporaries and successors argued that reproductively isolated breeding groups were the proper units of evolution by natural selection. In so doing, they provided the criteria for the currently traditional definition of 'species'; and they solved the problem of the evolution of unreal species, by explaining the sense in which evolving species are real.

THE SOLUTION

Before discussing the discovery of the role of breeding groups as units of evolutionary change, and the subsequent definition of 'species' as reproductively isolated breeding groups, it is important to consider why Darwin himself was not led to this species concept. Actually, a case has been made that Darwin held this very view of species in his transmutation notebooks of 1837–1839, and in his unpublished evolutionary essays of 1842 and 1844 (Kottler, 1979). In support of this thesis, historians have produced a number of quotes from these sources to show that Darwin emphasized reproductive isolation as a criterion for distinguishing species. For instance, early in his first notebook (July, 1837–February, 1838), Darwin insists that "A species as soon as once formed by separation or change in part of country, repugnance to intermarriage – settles it" (Darwin, 1960–1961, p. 24). And as he argues later in the first notebook,

... between species from moderately distant countries there is no test but generation (but experience according to each group) whether good species, and hence the importance naturalists attach to geographical range of species. (Darwin, 1960–1961, p. 212)

The interesting problem then becomes Darwin's apparent *abandonment* of the view of species to which his successors would return (Sulloway, 1979, p. 34).

But this may be a pseudoproblem, because the evidence in favor of Darwin's early adherence to the more modern view of species is not clear-cut. In the first place, Darwin's early emphasis on reproductive isolation is not at odds with traditional definitions of 'species' – definitions that emphasized reproductive criteria, but that also included constancy-of-type criteria. Remember Buffon's definition:

We should regard two animals as belonging to the same species if, by means of copulation, they can perpetuate themselves and preserve the likeness of the species.... (Buffon, 1749, p. 10; in Lovejoy, 1959, p. 93)

As it happens, there are passages in Darwin's notebooks and essays which suggest that, early on as well as later on in his career, he also considered constancy part of the meaning of 'species'. In the first notebook, we find:

Definition of species: one that remains at large with constant characters, together with other beings of very near structure.

The context of this definition suggests that, even when two species have similar characteristics and neighboring or overlapping ranges, reproductive isolation keeps them distinct by preventing the mixing of their distinguishing characteristics. But the definition also requires that the isolated species remain constant.

That Darwin considered constancy as well as reproductive criteria important in defining 'species', is also evident in his essay of 1844, where, in a discussion of the difficulty in distinguishing races from species, Darwin maintains,

... comparing, on the one hand, the several species of a genus, and on the other hand several domestic races from a common stock, we cannot discriminate them by the amount of external difference, but only, first, by domestic races not remaining so constant or being so 'true' as species are; and secondly by races always producing fertile offspring when crossed. (Darwin and Wallace, 1958, p. 243)

So it may be the case that Darwin never really gave up the more modern definition of 'species' in favor of the traditional definition. Rather, he may never have given up the traditional definition. To be sure, there is plenty of confusion about Darwin's view(s) of species. But wouldn't we expect Darwin to be just a little confused about the meaning of 'species', given that he was formulating a theory of the evolution of species, while at the same time the term 'species' had nonevolutionary connotations?

At any rate, although it is not clear that Darwin ever considered species as just reproductively isolated breeding groups, one of the reasons for thinking that he *might* have considered them as such is that, at least until 1844, Darwin placed great emphasis on the role of reproductive isolation in the divergent evolution of species (Kottler 1978, and Sulloway 1979). The importance of reproductive isolation for divergent evolution, and the manner in which an isolation theory of divergence supports the concept of a species as an isolated breeding group, should become clear in what follows. But Darwin's account of divergent evolution in 1859, in the *Origin*, placed less emphasis on the need for isolation; and to many of Darwin's contemporaries and successors, it appeared that his account of divergence actually discounted the need for isolation (Sulloway, 1979, pp. 41–60). In response to Darwin's account of divergence in the *Origin*, it was argued that natural selection changes a reproductively isolated breeding group *as a whole*, and that divergent evolution occurs only *between*, not within, such groups. This argument led to the recognition that isolated breeding groups are the proper units of evolutionary change, and thus led to the theory-laden redefinition of 'species' as reproductively isolated breeding groups. Thus it became possible for real species to evolve.[10]

According to Darwin's account in the *Origin* (Darwin, 1859, pp. 111–126), divergent evolution was initiated by the occurrence of variant individuals, whose particular variations allowed them to seek resources not utilized by the remainder of the species. Reasoning that the ability to avoid resource competition is beneficial, and that beneficial variations are preserved and accumulated by natural selection, Darwin concluded that the more divergent variants would be selected. The result of constant selection for the divergent variants of a group would be evolutionary divergence of type. As Darwin explained,

> . . . during the modification of the descendants of any one species, and during the incessant struggle of all species to increase in numbers, the more diversified these descendants become, the better will be their chance of succeeding in the battle of life. Thus the small differences distinguishing varieties of the same species will steadily tend to increase

until they come to equal the greater differences between species of the same genus, or even of distinct genera. (Darwin, 1859, p. 128)

Thus, Darwin considered natural selection alone sufficient to account for the branches of an evolutionary tree.[11]

G. J. Romanes, otherwise one of Darwin's strongest supporters, challenged Darwin's account of divergent evolution in 1886, on the grounds that the acknowledged means of inheritance (at the time) precluded such a simple explanation of divergence. In particular, inheritance considerations suggested that the reproductive isolation of the diverging types was an additional prerequisite for divergent evolution. According to the popular "blending theory of inheritance," characters manifested by parents were thought to be blended in their offspring (for instance, it was often noted mulattos are intermediate in color between their black and white parents). Thus Romanes argued that blending inheritance would produce *uniformity* of character among interbreeding individuals. *Divergent* traits could not be maintained over the course of generations if possessors of the different traits interbred and their offspring were intermediate in character. On the other hand, if two sections of an original breeding group were prevented from interbreeding, then character divergence between them would be possible. For in this case, as Romanes explained, "the two . . . divided sections of the species are free to develop independent histories without mutual intercrossing" (Romanes, 1886, p. 353). Romanes also spoke of reproductively isolated breeding groups as having "independent genetic histories" and "independent varietal histories". This 'independence' reflects the idea that two separated groups are no longer one evolutionary unit. Two such groups can accumulate and maintain different variations. Thus, in Romanes's work, we find a candidate for a proper unit of evolution by natural selection: namely, the breeding group.

The American naturalist J. T. Gulick also challenged Darwin's account of divergence, on the grounds that uniformity of character would ultimately prevail among interbreeding individuals. And, like Romanes, Gulick cited blending inheritance as a cause of this uniformity. But Gulick also emphasized another uniformity-producing factor. He argued that, over the course of generations, naturally selected traits would tend to spread throughout the range of a breeding group. That is, if an advantageous trait were to arise that somehow avoided being blended with other characters, the reproductive links between members of a breeding group would ensure nevertheless that the group became uniform with respect to the trait. Gulick emphasized both uniformity-producing factors in the following passage:

When there is free crossing between the families of one species, will not any peculiarity that appears in one family either be neutralized [i.e., by blending] by crosses with families possessing the opposite quality, or being preserved by natural selection, while the opposite quality is gradually excluded, will not the new quality gradually extend to all branches of the species, so that, in this way or in that, increasing divergence of form will be prevented? (Gulick, 1888, p. 282)

This new uniformity-producing factor is especially interesting because it continues to be applicable, even though blending inheritance has been replaced by Mendel's nonblending theory of inheritance. Indeed, modern evolutionists still emphasize the uniformity of character within breeding groups. For example, John Maynard Smith argues that,

... it is the free interbreeding within species and the absence of hybridization between them which are responsible for the relative uniformity of structure of members of a given species.... (Maynard Smith, 1958, p. 157)

At any rate, Gulick recognized that evolutionary divergence of character could not occur within one breeding group. For one reason or the other, a breeding group evolves uniformly – as a whole. Gulick referred to isolated breeding groups as "communities of evolution" – the sorts of entities that natural selection changes as a whole. This rather sophisticated population concept is articulated in the following passages:

As *community of evolution* arises where there is community of breeding between those that, through superior fitness, have the opportunity to propagate, so I believe it will be found that divergent evolution arises where there is separate breeding of the different classes of the successful. In other words, exclusive breeding of other than average forms causes monotypic evolution, and segregate breeding causes divergent or polytypic evolution. (Gulick, 1888, p. 284, my emphasis)

When separate generation comes in between two sections of a species they cease to be *one aggregate, subject to modification through the elimination of certain parts.* Both will be subject to similar forms of natural selection only so long as the circumstances of both and the variations of both are nearly the same, but they will no longer be members of *one body [within] which the selection process takes place.* (Gulick, 1888, pp. 312–313, my emphasis)

Implicit in Romanes' and Gulick's emendation of Darwinian evolutionary theory is the characterization of species as reproductively isolated breeding groups. For on the one hand, species are said to evolve by natural selection and to diverge evolutionarily from one another; and on the other hand, the proper units of evolution by natural selection and divergent evolution are characterized as isolated breeding groups. Yet neither Romanes nor Gulick ever explicitly defined 'species' in this manner. Nor did David Starr Jordan, another turn-of-the century advocate of the isolation theory of divergence. At least Jordan thought it worthwhile to address the problem of the evolution of unreal species, though, insisting that "in discussing the origin of species, we first premise that species in nature exist" (Jordan, 1916, p. 379); but he hesitated to define 'species' explicitly.[12]

Thus, though terms may eventually derive their meanings from the theories in which they are employed, it appears that they are not immediately theory laden. The thesis that terms derive their meanings from theories suggests that a term shared by successive, alternative theories will have different meanings. But that thesis alone does not suggest at what point in the history of the new theory the term in question is redefined. Apparently though, the ties of present terms with past opinion are not easily overcome. The past significantly constrains the future.

It is difficult to determine when definitions of 'species' as reproductively isolated breeding groups first gained anything like general acceptance. Of course, the theory must have gained general acceptance before the theory-laden definitions did. Ernst Mayr (1980) has suggested that the turn-of-the-century popularity of the mutation theory of evolution, an alternative to the amended Darwinian evolutionary theory, favored a rather different view of species. According to this theory, species are fairly constant during their lifetime; new species are formed rather instantaneously, when a distinct mutation occurs in a large number of individuals. To the extent that the mutated individuals are viable, fertile, and able to compete with the other forms in the area, they breed together and perpetuate the new type. This theory of natural history led its leading proponent, the Dutch geneticist Hugo de Vries, to a definition of 'species' very similar to the old one. In distinguishing species, de Vries maintained,

> Pedigree culture is the method required and any form which remains constant and distinct from its allies in the garden is to be considered an elementary species. (de Vries, 1905, p. 10)

Thus, definitions of 'species' in terms of reproductively isolated breeding groups awaited the resurrection of Darwinian evolutionary theory in the 20s and 30s. Following that period, the evolutionists Theodosius Dobzhansky, Ernst Mayr, and George Gaylord Simpson emerged as the leading spokesmen for the new 'species' definitions (see especially Dobzhansky, 1935, and 1937, pp. 303–321; Mayr, 1940, and 1942, pp. 102–146; and Simpson, 1951). For example, in his *Systematics and the Origin of Species*, published in 1942, Mayr defined 'species' as "groups of actually or potentially interbreeding natural populations, which are reproductively isolated from other such groups" (Mayr, 1942, p. 120). Note that Mayr's definition *reflects* the theoretical status of species as units of evolution, but without *explicitly stipulating* that species are evolutionary entities. In other words, Mayr's definition *allows* species to be evolutionary entities, without *requiring* as much. Simpson's later definition is more theory laden in this regard. For, according to Simpson, 'species' is more appropriately defined as "a phyletic lineage (ancestral-descendant sequence of interbreeding populations) evolving independently of others, with its own separate and unitary evolutionary role . . . " (Simpson, 1951, p. 289; see also Simpson, 1961, p. 153). At any rate, thanks in large part to Dobzhansky's, Mayr's, and Simpson's support, the new 'species' definitions gained increasing acceptance. Following theory change, then, theory-laden definitions are replaced by theory-laden redefinitions. The process is complete – at least for the time being.[13]

A final point before concluding concerns a remark made by Mayr in the presentation of his 'species' definition. As a preface to his definition, he suggested that questions about the origin and evolution of species could not properly be discussed until a definition of 'species' had been provided (Mayr, 1942, p. 114). In keeping with this maxim, his definition preceded his discussion of the origin and evolution of species in the work in question. Actually, the title of that work, *Systematics and the Origin of Species*, reflects the order and strategy of presentation. But however reasonable this order is from a pedagogical point of view, it is somewhat misleading from a historical point of view. It is the thesis of this essay that certain terms derive their meanings from the theories in which they are couched, and that the modern definition of 'species' was not the historical *prerequisite* for, but the *result* of, the formulation and acceptance of modern evolutionary theory.

CONCLUSION: ANOTHER PROBLEM

Summing up, then, what's in a word? Often, scientific terms connote theories; and this makes it difficult to express a new, alternative theory in the same terms as the old theory it is supposed to replace. This tendency to preserve the past can be overcome by simply replacing the terminology of the past with new terminology that has no ties to past opinion. Or, proponents of alternative theories may retain, but redefine, past terminology in such a way that they can express their alternatives without contradicting themselves or introducing conceptual confusions.

These considerations shed some light on a conceptual confusion implicit in Darwin's theory of the evolution of species. Given the loaded meaning of the term 'species' derived from nonevolutionary theories of natural history, Darwin was led to question the reality of species. For to admit that species were real, according to the old definition, was to admit that species were constant. Darwin might have replaced the term 'species', as Lavoisier and others replaced the term 'phlogiston' when it was discovered that nothing existed corresponding to the traditional definition of that term. But Darwin did not abolish the term 'species'. Nor did he redefine the term in order to allow for the reality of mutable species. Thus arose the conceptual difficulty concerning the evolution of unreal species. That evolutionary theory is no longer plagued by this difficulty, is due to the fact that Darwin's successors redefined 'species' in terms of evolutionary theory, thus explaining the new sense in which species are real.

I wish that, as a means of concluding, I could simply sum up the various aspects of this analysis of the Darwinian Revolution. But in all honesty, a frustrating and difficult question remains. I have rather strategically avoided the question until now, even though it has bothered me throughout. It concerns the difficulty of making a rational choice between alternatives in a scientific revolution. Thomas Kuhn has characterized scientific revolutions in terms of this very issue, so a prefatory quote from Kuhn is in order:

> The inevitable result [of a scientific revolution] is what we must call, though the term is not quite right, a misunderstanding between the two competing schools. The laymen who scoffed at Einstein's general theory of relativity because space could not be "curved" – it was not that sort of thing – were not simply wrong or mistaken. Nor were the mathematicians, physicists, and philosophers who tried to develop a Euclidean version of Einstein's theory. What had previously been meant by space was necessarily flat, homogeneous, isotropic, and unaffected by the presence of matter. If it had not been, Newtonian physics would not have worked. (Kuhn, 1970, p. 149)

My question is this. Isn't it also possible that the nonevolutionists were not simply wrong or mistaken about the immutability of species? That's a difficult question, when you consider what's in a word.

Harvard University

NOTES

I am very grateful for the help of my friends and colleagues Fred Churchill, Jonathan Hodge, David Hull, Philip Kitcher, David Kohn, Ernst Mayr, Jane Maienschein, James

Paradis, Michael Ruse, Sam Schweber, Peter Stevens, and Frank Sulloway. This paper will also appear in *Journal of the History of Biology* 15 (1982): 215–239.

[1] Ernst Mayr has remarked, concerning Darwin's denial of the reality of species, that, "Having thus eliminated the species as a concrete unit of nature, Darwin had also neatly eliminated the problem of the multiplication of species. This explains why he made no effort in his classical work to solve the problem of speciation" (Mayr, 1957, p. 4). However, one could also say that Darwin neatly eliminated the problem of the evolution of species as well, by denying the reality of species. Yet Darwin made a significant case for the evolution of species. Moreover, it is not quite true that Darwin made no effort in the *Origin* to solve the problem of speciation. As faulty an account as it may have been, his theory of divergent evolution was supposed to solve that problem.

[2] Several general, historical accounts of the connections between nonevolutionary theories of natural history and static species concepts are available. See in particular the surveys by David Hull (1967) and Ernst Mayr (1957). Though connections between theories of natural history and 'species' definitions are drawn in these accounts, however, the connections are not explained in terms of the general connection between theory and definition – i.e., in terms of theory-laden meaning.

[3] As Phillip Sloan (1979) has persuasively argued, Buffon's definition of "species" is part of a tradition of historical-genealogical definitions – a tradition further articulated by Kant and carried on by Johann Karl Illiger. Sloan quite properly contrasts this tradition with one in which species were viewed less historically, more as morphological kinds. The last-mentioned tradition was perhaps most famously represented by Linnaeus; as he defined "species" in his *Philosophia Botanica*, "Species are as many as there were diverse forms produced by the Infinite Being; which forms according to the appointed laws of generation, produce more individuals but always like themselves" (1751: 99; translated by Ramsbottom 1938: 196).

Without contesting the reality of differences in *emphasis* on historical genealogies vs. nonhistorical morphological kinds, it is nevertheless important as well not to overlook the extent to which Buffon's conception is still tied to the notion of species-qua-kind, and the extent to which Linnaeus's conception includes the notion of species-qua-lineage. After all, Buffon stresses not only that members of a species "perpetuate themselves," but also that they "preserve the likeness of the species." And Linnaeus stresses that members of a species "produce more individuals," as well as stressing that those produced are "always like themselves."

[4] It seems that Lamarck was responsible for the same conceptual confusion that Agassiz recognized in Darwin's work – namely, a theory of the evolution of unreal species. It may be that Agassiz also recognized the confusion in Lamarck's work, since he refers, nonspecifically, to "much confusion of ideas in the general statement, of the evolution of species, so often repeated of late." Whether this confusion had the same source in Lamarck's thought, as I argue it had in Darwin's thought, remains to be explored. At any rate, the fact that Lamarck got away with it (Lyell did not accuse him of utter confusion, for instance) may have reinforced Darwin's thinking along these lines.

[5] The Swiss-American naturalist Louis Agassiz also insisted on using the term 'species' in such a way as to make the evolutionists' job difficult, if not impossible. But the connotations that he attached to the term were quite different from the nonevolutionary connotations considered thus far. According to Agassiz, species were categories of thought in the mind of God – the Creator's plans: "those [classificatory] systems to

which we have given the names of the great leaders of our science who first proposed them being in truth but translations into human language of the thoughts of the Creator" (Agassiz, 1857, p. 9). As such, species were real, but not materially real. The difference between this view and the view of species as genealogies of nonvarying organisms is clear from the following passage: "When first created, animals of the same species paired because they were made for one another; they did not take on another in order to build up their species, which had full existence before the first individual produced by sexual connection was born" (Agassiz, 1857, p. 173). Employing this concept of species, Agassiz argued that the evolution of species and their common descent were impossible. Since species were only intellectual entities with only intellectual connections, they could not possibly vary materially or have genealogical connections: "As the community of characters among the beings belonging to these different categories arises from the intellectual connection which shows them to be categories of thought, they cannot be the result of a gradual material differentiation of the objects themselves" (Agassiz, 1860, p. 89). The theoretical basis of Agassiz's species concept, and the constraints it placed on discussions of the evolution of species are clear enough. But see Asa Gray's response in note 7.

[6] As I hinted in the introduction, theory ladenness is a much-discussed notion in contemporary philosophy of science. It is also a much-embattled notion. So perhaps I should make my nonallegiances clear. I am not attempting to solve "the" problem of theory ladenness here, but only to point out how I think such problems entered into the resolution (or dissolution – see my conclusion) of a scientific controversy. What I mean by "theory ladenness" is goint to be spelled out only, or mostly, in terms of 19th century discussions of that notion. Thus, I will concentrate mainly on the problem of how *definitions* of scientific terms reflect the theories in which they are employed. Philosophers who stress this sort of meaning determination are also apt to stress the difficulty of formulating new theories in old terms. Philosophers have also stressed a different sort of meaning determination, however – namely, meaning as reference. Among the last-mentioned philosophers are some who argue that communication over the course of a scientific dispute is not as difficult as it might at first appear. Communication is possible because the terms whose definitions are in question are nonetheless used by all disputants to refer to the same things in the world. Israel Scheffler (1967) is among the proponents of this solution. Philip Kitcher's defense (1978), which is updated in terms of recent developments in semantics, deals with the phlogiston case in particular, and is highly recommended. That approach to the Darwinian revolution is, however, grist for another mill (soon to be ground).

[7] Darwin did not face Agassiz's objection alone. He also acknowledged Asa Gray's defense: " . . . it [Agassiz's review] hardly seems worth a detailed answer (even if I could do it, and I much doubt whether I possess your skill in picking out salient points and driving a nail into them), and indeed you have already answered several points" (to Asa Gray, August 11, 1860, in F. Darwin, 1887, II, p. 124). Gray's response to Agassiz's charge of conceptual confusion is, well, interesting. Following Agassiz (see note 5), he suggests distinguishing between species as real material entities, and species as real categories of thought. And he interprets Darwin's position on the reality of species as a denial of their material reality, but an affirmation of their reality as categories of thought. Thus he sidesteps Agassiz's objection that, "if species do not exist *at all* [my emphasis], as the supporters of the transmutation theory maintain, how can they vary?"

Gray then had to consider Agassiz's argument that species, as categories of thought, could not possibly vary. Gray changed the objection somewhat in responding to it: from the original problem of how species as intellectual entities can vary materially, to the problem of how species as intellectual entities can vary at all. In response to the latter problem, he suggested that ideas can change in time in the mind of the beholder. Of course, in Gray's scheme of things, God was the beholder. The evolution of intellectual species thus amounted to something like God's changing his mind. Speaking of "Divine thoughts," Gray argued, "allowing that what has no material existence can have had no material connection or variation, we should yet infer that what has intellectual existence and connection might have intellectual variation" (Gray, 1860, in 1963, p. 137). As clever a response as this is, it is certainly not a defense of Darwin's theory.

8 In response to Cuvier's definition of 'species', Haeckel argued, "In a closer examination of this definition . . . , it becomes at once evident that it is neither theoretically satisfactory nor practically applicable. Cuvier, with this definition, began to move in the same circle in which almost all subsequent definitions of species have moved, through the assumption of their immutability" (Haeckel, 1879, p. 51). But Haeckel was so convinced that a nonevolutionist's definition of 'species' would be inadequate, that he did not explicitly point out the inadequacies of Cuvier's definition. We can hazard a guess, though, as to what he might have had in mind.

Haeckel presented the following as Cuvier's definition of 'species': "All those individual animals and plants belong to one species which can be proved to be descended from one another, or from common ancestors, or which are as similar to these as the latter are among themselves" (Haeckel, 1879, pp. 50–51; and see Cuvier, 181, pp. 119–120; Whewell also adopted Cuvier's definition, 1847, I, p. 505). On this definition, generations of one lineage, no matter how different, are generations of the same species. So this definition at least allows the possibility of the evolution of species. However, it does not seem possible, on this definition, for one species to be the descendant of another. For on this definition, *all* a parent's descendants are members of the parent's species. So the definition rules out the common-descent account of the origin of species.

9 In the 1872 edition of *Principles*, Lyell also pointed out that Lamarck had not only questioned prevailing theories of the immutability of species, but had also proposed a suitable redefinition of 'species'. Following the nonevolutionary definition of 'species' he had cited in earlier editions, Lyell added, "Lamarck proposed, therefore, to amplify the received definition in the following manner. 'A species consists of a collection of individuals resembling each other, and reproducing their like by generation, *so long as* [my emphasis] the surrounding conditions do not alter to such an extent as to cause their habits, characters, and forms to vary'" (Lyell, 1872, p. 249).

10 The late nineteenth-century controversies concerning the role of isolation in divergent evolution constitute a rich segment of the history of evolutionary theory – too rich a segment to be treated adequately here. The German naturalist Moritz Wagner first challenged Darwin's account of divergence in 1868. But his arguments drew only slightly compromising recognition from Darwin in the fifth and sixth editions of the *Origin*. G. J. Romanes, in 1886, and J. T. Gulick, in 1888, reopened the controversy with independently conceived arguments against Darwin's theory of divergence. Their arguments evoked substantial response, favorable and unfavorable, from the leading evolutionary thinkers of the day (e.g., A. R. Wallace 1889, pp. 142–151, 180–184). Only Romanes's and Gulick's objections to Darwin's account are considered in this essay. Romanes's

final position on the role of reproductive isolation in divergent evolution is the topic of the third volume of his *Darwin and After Darwin* (1897). Gulick's *Evolution, Racial and Habitudinal* (1905) is essentially a composite of his earlier works on the isolation theory of divergent evolution.

A number of very good secondary accounts of the isolation controversies are now available, beginning, in the recent past, with Ernst Mayr's (1959), and ending most recently with the exceptional analysis of Mayr's student, Frank Sulloway (1979). Also very helpful are the accounts of Peter Vorzimmer (1970, pp. 159–185), Jack Lesch (1975), and Malcolm Kottler (1978). The account presented in this essay differs from the accounts above in its emphasis on the discovery of the proper units of evolutionary change, rather than just the discovery of the role of isolation in divergent evolution. Ghiselin (1969, pp. 89–102) attributes to Darwin a more up-to-date appreciation of the status of species as units of evolution than I am willing to grant.

[11] Frank Sulloway argues convincingly that Darwin's theories of divergence changed from more isolationist (in emphasis) to less isolationist as Darwin took into account 1) evidence for sympatric speciation in plants, and 2) the problem of speciation in large continental areas that were ecologically partitioned but not subdivided by geographically isolating barriers (Sulloway, 1979, pp. 39–45). Sam Schweber also sees the principle of divergence as a solution to problems posed by apparent sympatric speciation in plants, and speciation in large areas like bodies of water that are not clearly geographically subdivided (Schweber, 1980, p. 209). But Schweber is concerned to substantiate quite a different sort of inspiration as well: namely British political economics. That is, as Schweber suggests, Darwin's principle of divergence is more than coincidentally similar to Adam Smith's theory of the division of labor, according to which more specialized laborers better sustain themselves by more readily locating niches in the work force (Schweber, 1980, especially pp. 257–275).

[12] Although Jordan declined to define 'species', he characterized them in a manner which shows the long-lasting influence of the traditional definition: "It [a species] is merely one particular crowd or mass of living things, giving rise by processes of reproduction to a succession of similar organisms, not all alike but nearly alike, so that for ordinary scientific purposes one name may serve for all" (Jordan, 1916, p. 379).

[13] Mayr presumably had Simpson's definition in mind when he argued that 'species' need not be defined explicitly as evolutionary entities. The evolution of species is a fact, and thus is a superfluous addition to the definition: "species are evolved and evolving. Again this is true from the individual to the highest categories and adds nothing to the definition" (Simpson, 1957, p. 18). Perhaps Mayr's definitional frugality is a more reasonable *prescription* for definitional practice, though, then it is an accurate *description* of that practice.

REFERENCES

Agassiz, L. 1857, *Contributions to the Natural History of the United States of America*, Volume 1: *Essay on Classification*, Little, Brown, Boston. Reprinted by Harvard University Press, Cambridge, 1962.

Agassiz, L.: 1860, *Contributions to the Natural History of the United States of America*, Volume 3, Little, Brown, Boston.

Buffon, G.: 1749, *Histoire Naturelle*, Volume 2, Imprimerie Royale, Paris.

Cohen, I. B.: 1976, 'William Whewell and the Concept of Scientific Revolution', in R. S. Cohen, *et al.* (eds.), *Essays in Memory of Imre Lakatos*, D. Reidel, Dordrecht, Holland. pp. 55–63.

Crosland, M.: 1962, *Historical Studies in the Language of Chemistry*, Dover, New York.

Cuvier, G.: 1818, *Essay on the Theory of the Earth*, Kirk and Mercein, New York.

Darwin, C.: 1859, *On the Origin of Species*, Murray, London. Facsimile of the first edition by Harvard University Press, Cambridge, 1967.

Darwin, C.: 1960–1961, 'Darwin's Notebooks on the Transmutation of Species', Parts I–V, *Bulletin of the British Museum (Natural History), Historical Series 2.*

Darwin, C. and Wallace, A. R.: 1958, *Evolution by Natural Selection*, Cambridge University Press, Cambridge. Includes Darwin's essays of 1842 and 1844.

Darwin, F. (ed.): 1887, *The Life and Letters of Charles Darwin, Including an Autobiographical Chapter*, 3 volumes, Murray, London.

Darwin, F. and Seward, A. C. (eds.): 1903, *More Letters of Charles Darwin*, 2 volumes, Murray, London.

de Vries, H.: 1905, *Species and Varieties, Their Origin by Mutation*, Open Court, Chicago.

Dobzhansky, T.: 1935, 'A Critique of the Species Concept in Biology', *Philosophy of Science* 2, 344–355.

Dobzhansky, T.: 1937, *Genetics and the Origin of Species*, Columbia University Press, New York.

Ghiselin, M.: 1969, *The Triumph of the Darwinian Method*, University of California Press, Berkeley.

Gray, A.: 1860, 'Darwin and his Reviewers', *Atlantic Monthly* 6, 406–425. Reprinted as Part III of 'Natural Selection not Inconsistent with Natural Theology', *Darwiniana*, 106–145, originally published by Appleton, New York, 1876, and reprinted by Harvard University Press, Cambridge, 1963.

Gulick, J. T.: 1888, 'Divergent Evolution through Cumulative Segregation', *Journal of the Linnean Society, Zoology* 20, 189–274. Reprinted in *Annual Report of the Smithsonian Institution*, Government Printing Office, Washington, 1893.

Gulick, J. T.: 1905, *Evolution, Racial and Habitudinal*, Carnegie Institution, Washington.

Haeckel, E.: 1879, *The History of Creation*, 2 volumes, Appleton, New York.

Hull, D.: 1967, 'The Metaphysics of Evolution', *British Journal for History of Science* 3, 241–268.

Jordan, D. S.: 1916, 'The Laws of Species Forming', *American Museum Journal (Natural History)* 16, 379–380.

Kitcher, P.: 1978, 'Theories, Theorists, and Theory Change', *Philosophical Review* 87, 519–547.

Kottler, M.: 1978, 'Charles Darwin's Biological Species Concept and the Theory of Geographic Speciation: The Transmutation Notebooks', *Annals of Science* 35, 275–297.

Kuhn, T. S.: 1970, *Structure of Scientific Revolutions* (2nd ed.), University of Chicago Press, Chicago.

Lesch, J.: 1975, 'The Role of Isolation in Evolution: George John Romanes and John T. Gulick', *Isis* 66, 483–503.

Linnaeus, C.: 1751, *Philosophia Botanica*, Kiesewetter, Stockholm.

Lovejoy, A.: 1959, 'Buffon and the Concept of Species', in B. Glass (ed.), *Forerunners of Darwin*, 84–113, Johns Hopkins University Press, Baltimore.

Lyell, C.: 1835, *Principles of Geology*, 4 volumes, Murray, London.

Lyell, C.: 1872, *Principles of Geology*, 2 volumes, Murray, London.

Maynard Smith, J.: 1958, *The Theory of Evolution*, Penguin, Baltimore.

Mayr, E.: 1940, 'Speciation Phenomena in Birds', *American Naturalist* 74, 249–278.

Mayr, E.: 1942, *Systematics and the Origin of Species*, Columbia University Press, New York.

Mayr, E. *et al.*: 1953, *Methods and Principles of Systematic Zoology*. McGraw-Hill, New York.

Mayr, E.: 1957, 'Species Concepts and Definitions', in E. Mayr (ed.), *The Species Problem*, American Association for the Advancement of Science, Washington.

Mayr, E.: 1959, 'Isolation as an Evolutionary Factor', *Proceedings of the American Philosophical Society* 103, 221–230.

Mayr, E.: 1972, 'The Nture of the Darwinian Revolution', *Science* 176, 981–989.

Mayr, E.: 1980, 'The Role of Systematics in the Evolutionary Synthesis', in E. Mayr and W. Provine (eds.), *The Evolutionary Synthesis*, Harvard University Press, Cambridge.

Poulton, E. B.: 1903, 'What is a Species?', *Proceedings of the Entomological Society of London*, lxxvii-cxvi.

Ramsbottom, J.: 1938, 'Linnaeus and the Species Concept', *Proceedings of the Linnean Society of London* 150, 192–219.

Romanes, G. J.: 1888, 'Physiological Selection; and Additional Suggestion on the Origin of Species', *Journal of the Linnean Society, Zoology* 19, 337–411.

Romanes, G. J.: 1897, *Darwin and After Darwin*, Volume 3: *Post-Darwinian Questions*, Open Court, Chicago.

Scheffler, I.: 1967, *Science and Subjectivity*, Bobbs-Merrill, Indianapolis.

Schweber, S.: 1980, 'Darwin and the Political Economists: Divergence of Character', *Journal of the History of Biology* 13, 195–289.

Simpson, G. G.: 1951, 'The Species Concept', *Evolution* 5, 285–298.

Simpson, G. G.: 1961, *Principles of Animal Taxonomy*, Columbia University Press, New York.

Seebohm, H.: 1883, *A History of British Birds*, Porter, London.

Sloan, P. R.: 1979, 'Buffon, German Biology, and the Historical Interpretation of Species', *British Journal for the History of Science* 12, 109–153.

Sulloway, F.: 1979, 'Geographic Isolation in Darwin's Thinking: The Vicissitudes of a Crucial Idea', *Studies in the History of Biology* 3, 23–65.

Vorzimmer, P.: 1970, *Charles Darwin: The Years of Controversy*, Temple University Press, Philadelphia.

Wagner, M.: 1868, *Die Darwin'sche Theory und das Migrationgesetz des Organismen*, Leipzig: Duncker und Humblot. Translated by J. Laird as *The Darwinian Theory and the Law of the Migration of Organisms*, Stanford, London, 1873.

Wallace, A. R.: 1889, *Darwinism*, Macmillan, London.

Whewell, W.: 1837, *History of the Inductive Sciences*, 3 volumes, Parker, London.

Whewell, W.: 1847, *Philosophy of the Inductive Sciences*, 2 volumes, Parker, London.

DAVID HULL

COMMENTS ON BEATTY

I find myself, as a commentator on John Beatty's paper, in the unhappy position of agreeing with everything he says both great and small. My only recourse is to expand on what he has said, expressing myself in certain instances somewhat more polemically and less judiciously.

Beatty's thesis is that some of the confusion which greeted Darwin's *Origin of Species* resulted from his retaining the old term 'species' even though species for Darwin were not exactly what they had been for earlier biologists. Beatty sees this state of affairs as a good example of the problems which arise because of the theory-ladenness of scientific terms. However, before embarking on this general thesis, Beatty feels called upon to exorcise the evil spirits of present-day historiography of science. Finding a 19th-century example of a present-day view about scientific change might seem to some as reading the present into the past. For example, in contrasting the philosopher with the historian, I. Bernard Cohen (1977: 345) makes the following remarks:

> We may see in this episode why the historian feels strongly that our modern methods of mathematical or logical analysis should never place a screen between us as observers and the historical conditions of discovery. Indeed, to many historians, the major danger in the writing of history by nonhistorians (and even by some members of the profession) is the anachronistic application of our present canons of logic and mathematics and of scientific knowledge to prior experiments, laws, and theories.

I do not mean to make light of the invidious influence of presentism in the writing of history, especially history of science. Even after the dangers of presentism are pointed out, the ease with which we can impart anachronistic views to our historical subjects is really frustrating. However, I think present-day historians sometimes allow their fear of presentism to drive them to unnecessary extremes. Like it or not, we are lodged squarely in the present. Pretending otherwise can lead to nothing but confusion. The four areas in which presentism has presented the greatest difficulties are in matters of truth, reasoning, morals and meaning (Hull, 1979).

When Europeans embarked on their long voyages of discovery, they did not know about vitamins. Today we do. If an historian discovers that the

Michael Ruse (ed.), Nature Animated, 101–108.

sailors from a particular country suffered from scurvy, while those from another country did not, he might well look to the diets of these sailors to explain the difference – even though no one at the time might have suspected diet as an explanation. Perhaps in biblical times, people believed in miracles. Some still do today. But no present-day historian is about to explain any phenomenon in terms of miracles, even if it occurred in biblical times. Stating that the people at the time believed that Jesus was raised from the dead is not quite the same thing as stating that he was. Nor does it help to retreat from making claims about what actually happened to people's beliefs. Beliefs are no easier to substantiate than any other empirical phenomena, usually harder. The reticence which historians have about stating that something actually happened has nothing in particular to do with history but with traditional philosophical problems involved with belief, truth and knowledge. The appropriate response to someone who says that we are led to believe what we believe by socioeconomic forces is, were you lead to this belief by socioeconomic forces?

Our views about how to go from some knowledge to more knowledge have changed through the years. Although no one, including J. S. Mill, was as much of an 'inductivist' as philosophical rational reconstructions of this position might lead one to believe, scientists in 19th-century England claimed to place a greater emphasis on evidence than present-day philosophers argue is appropriate. In the investigation of a period in history, should an historian use the methods of investigation common in the period under study or his own? The answer to this question not only seems obvious, it is obvious. There is something desperately wrong with a principle of historiography which requires us to write bad history. At one time, historians happily interpreted past ages in their own image. Today we try not to, even when we are writing the history of these early histories. Why should someone who claims that reason, argument and evidence really have very little, if anything to do with people making up their minds present reasons, arguments and evidence for this claim? As W. R. Albury remarks in connection with the paper by Camille Limoges, the only answer consistent with this claim is distressingly cynical.

Was Aristotle a pragmatist? Was he a Quaker or antiflouridation? Such questions are clearly guilty of presentism of the most blatant sort. An author would find great difficulty in getting papers which addressed these issues published in most history journals, but strangely enough, present-day historians who would never dream of condemning Aristotle for believing in the immutability of species are quick to condemn him for being sexist. The moral standards used to make these decisions are not those common at the time but those of the historian writing the piece. I think that noting differences in morals is good historiography, e.g., warning the reader that an act considered perfectly acceptable today was held in profound contempt during the period under investigation. I think that it is also within an historian's role to note departures from the moral standards of the day. But if an historian is to avoid "presentism" of the most blatant sort, he should try to neutralize his disdain for an historical figure because he behaved in ways incompatible with the historian's own moral standards. If he is consistent, he should also do his best to hide his admiration for an historical figure because he behaved in ways compatible with the historian's own moral standards. Conversely, if it is all right to applaud an historical figure one finds admirable, then it should be equally appropriate to condemn him in the opposite situation. (For two answers to the question, "Was Aristotle a sexist?" see Horowitz, 1976; and Morsink, 1979.)

I finally turn to the sort of relativism with which Beatty deals – the relativity of meaning. Considerable disagreement exists over the relativity of morals, much less over the relativity of truth and principles of good reasoning. With a few notable exceptions, nearly all philosophers agree that terms change their meanings. Do meteorologists really think that meteors affect the weather? Is a melancholic lover actually afflicted with too much black bile? Did Mendel really contribute anything to Mendelian genetics? The fact that words change their meanings presents real difficulties to the historian. Many of the most informative records of the past are couched in language, and language is tightly intertwined with the society in which it is used. A sexist society is likely to produce a sexist language, and conversely the sexism buried in language can subtly perpetuate it. An historian studying a culture and its language must come to understand both without being taken in by either. Early Mendelians called themselves 'Mendelians', claimed to trace their basic principles to Mendel, etc. However, none of this entails that Mendel and his writings actually played the roles attributed to them by the Mendelians. (For a discussion of this point, see Olby, 1979; and Brannigan, 1979.)

Today certain philosophers argue that scientific terms are to varying degrees theory-laden. Whether right or wrong in this connection, they are perfectly within their rights to attempt to test their views by studying actual episodes in science, even though the scientists at the time may never have thought of the issue or may have disagreed with it if they did. After all, few of the scientists investigated by present-day anti-presentist historians ever heard of presentism. Cohen (1976: 53) complains that current views of science are being "imposed harshly and anachronistically on the scientific events of the past". If by "harshly imposed", he means that our present-day views on the nature of science cause us to claim that Galileo said, thought, did, etc. what he did not say, think, or do, then I agree with him, but I do not see what "anachronistically" has to do with it. It would be just as wrong to impose harshly the views of science commonly held in Galileo's day on Galileo if he did not hold them. Conversely, I do not know what other ideas we are supposed to bring with us to the study of the past than our own. As anachronistic as they may be, they are all we've got.

Beatty's concern, however, is not the theory-ladenness of our meta-conceptions, but the theory-ladenness of such scientific terms as "species". A prevalent view about biological species from Aristotle to Darwin was that they are immutable. In fact, immutability was so central to the notion of species – all species and not just biological species – that anything which could undergo essential change could not count as a species. As O. A. Brownson (1884, 9: 491) was moved to remark, "The *differentia* of man, not being in the ape, cannot be obtained from the ape by development. This sufficiently refutes Darwin's whole theory."

According to traditional metaphysics, species have two characteristics in addition to mutability. They are also eternal and discrete. Given these three variables, eight different permutations are possible (see Figure 1).

eternal	*immutable*	*discrete*
yes	yes	yes
yes	yes	no
yes	no	yes
yes	no	no
no	yes	yes
no	yes	no
no	no	yes
no	no	no

Figure 1. Eight possible permutations of the three traditional attributes of natural kinds.

An interesting exercise might be to discover if at least one figure in the history of science could be found for each slot. In any case, the commonest view about species was that they are eternal, immutable and discrete. Using present-day terminology, two senses of each of these attributes can be distinguished – extensional and "metaphysical". To say that *Homo sapiens* is eternal from an extensional point of view means that at all times people exist in space and time. I take it that this was Aristotle's position. But many philosophers had a more 'metaphysical' notion in mind, e.g., individual people might cease to exist without peoplehood or personhood ceasing to exist.

Aristotle believed that species are immutable, but as numerous scholars have pointed out, he also believed that particular organisms can change their species, i.e., either produce or themselves become organisms belonging to a different species. All that Aristotle precluded was organisms doing so regularly and wholesale (Hull, 1967). As I understand Lamarck, he thought that the borderlines between species are not 'real' but that the order of species in several Great Escalators of Being is not arbitrary. As organisms move up the tree of life through successive generations, the branches of these trees of life stay the same.

Throughout the history of natural history, students of the living world – anti-presentism forbids my calling them biologists until 1802 – were aware of the variability which is so characteristic of organic species. However, throughout this history, these same workers maintained that species are discrete. But this they meant that characteristics could be divided into accidental characteristics which vary and essential characteristics which do not. Any organism deficient in one of its essential characteristics is a monster. Even if one could align organisms in a continuum, or several intersecting continua, the metaphysical correlates of species were believed to remain as discrete as triangles and quadrilaterals.

Darwin's view that species evolve gradually over long periods of time so that certain species go extinct permanently, either by ceasing to exist altogether or by evolving into genuinely new species, threatened all three of the preceding tenets. A metaphysically-inclined biologists might admit that species could go extinct, just as all the gold in the universe might cease to exist, but that in some sense the species as a metaphysical entity still existed. If the appropriate organisms were to reappear, the species would re-evolve. Darwin disagreed. Even though gold atoms might come into existence again after all gold atoms had ceased to exist, biological species, once extinct, cannot come into existence again. Biological species as segments of the phylogenetic tree are spatiotemporally localized (Ghiselin, 1974; Hull, 1976, 1978, 1980).

Darwin also maintained that new species arise, not by a single organism leaping the boundary between species, but by numerous organisms changing gradually and wholesale through time. According to Darwin, species are mutable in the most extreme sense of this term. If one follows species through time, the boundaries between them are not just fuzzy, they are non-existent. Of course, not all of Darwin's fellow evolutionists agreed with him on all counts. Huxley, for example, maintained that evolution is saltative. Species may not be eternal and immutable, but at least they are discrete.

Did Darwin think that in defining 'species' constancy of character was basic, reproductive continuity, both, first one and then the other? Beatty argues that no simple answer exists for this question, and I wholeheartedly agree. If Darwin was perfectly clear and consistent on this issue, he was the first biologist to be so and close to the last. Present-day biologists still not only disagree on these issues but also frequently confuse them. (And philosophers are no better; see Kitts and Kitts 1979; Caplan, 1980, 1981.) Why? I think the answer to this question requires an extension of Beatty's notion of theory-ladenness from the meanings of scientific terms to the ontological status of theoretical entities. Scientific theories not only influence the meanings of their constituent terms but also determine the metaphysical categories of the entities to which they refer.

According to early workers (and most present-day workers as well), species are natural kinds. As natural kinds, characteristics like being eternal, immutable and discrete are appropriate to them. Triangularity and quadrilaterality are excellent examples of natural kinds. One might reshape a wire triangle into a quadrilateral, but what would it mean for triangularity to evolve into quadrilaterality? If biological species are natural kinds, one species evolving into another should seem just as incomprehensible. It is one just that one disagrees with the claim but that one cannot even conceive of what the speaker intends. Lead and gold are also excellent examples of natural kinds. I know what it would mean to transmute a sample of lead into a sample of gold. In fact, it has been done. But I do not understand what one might mean by the claim that leadness can be transmuted into goldness. Natural kinds are simply not the sort of thing that can evolve. For example, Woodger's paradox, as Lindenmayer describes it, results from treating species inappropriately as sets. If species are sets, then gradual evolution is impossible. Lindenmayer opts for this conclusion. I prefer the opposite line of reasoning: since evolution is at least sometimes gradual, species are not sets.

According to Darwin, species are the sort of thing which can be eternal, immutable and discrete; they just happen not to be. (See Hodge's discussion of the serious consideration which Darwin gave to Brocchi's suggestion that species might have predetermined life spans the way that organisms do.) Even though biological species lack all the characteristics usually attributed to natural kinds, Darwin and everyone else still attempted to view them as natural kinds, albeit very peculiar natural kinds. I think that it was this that made it impossible for Darwin to come up with an appropriate definition of "species". If species are natural kinds, then constancy of character of some sort must matter. But reproductive continuity also seems to be relevant. What should one do when the two do not covary? Which should take priority? If constancy of character of some sort, then species are natural kinds and cannot evolve. If reproductive continuity, then species can evolve, but they are no longer appropriately viewed as natural kinds. They belong to quite a different category – not secondary substance but primary substance, not class but individual, not universal but particular.

Beatty concludes his paper with the question, "Isn't it possible, then, that the non-evolutionists were not simply wrong or mistaken about the immutability of species?" Beatty's implied answer to this question is that the evolutionists and non-evolutionists were not disagreeing over matters of fact but over definitions. The evolutionists and non-evolutionists meant something quite different by the term 'species'. Hence, the evolutionists' claim that species evolve could not contradict the non-evolutionists' claim that they do not. In my comments, I responded that it does not make any difference which way one puts it. If by "species" both sides meant those things commonly referred to as species at the time, then Darwin was claiming that non-evolutionists were wrong about the facts. If the non-evolutionists intended to include immutability in their definition of 'species', then the appropriate response is that species as non-evolutionists view them do not exist (see Beatty's postscript).

Early opponents of Darwin argued that, if species are natural kinds, they cannot evolve. The Darwinians countered that species, though natural kinds, can nevertheless evolve. On this point, I agree with Darwin's opponents. Their argument is perfectly cogent. They merely opted for the wrong conclusion. They concluded that species do not evolve, while I think that the appropriate conclusion is that they are not natural kinds. Instead they are historical entities, spatiotemporally localized entities, exactly the sort of thing which one might expect to be temporary, changeable and relatively indistinct, and not eternal, immutable and absolutely discrete. The sort of subtle change in meaning which takes place in conceptual evolution frequently causes considerable confusion. Retaining the same term for an entity when it is reinterpreted as belonging to a different metaphysical category is guaranteed to produce even greater confusion. The fact that this particular issue is being raised only now, more than a century after the publication of the *Origin of Species*, indicates exactly how deeply metaphysical beliefs run.

University of Wisconsin-Milwaukee

REFERENCES

Brannigan, A.: 1979, 'The Reification of Mendel', *Social Studies of Science* 9, 423–454.

Browson, O. A.: 1884, *The Works of Orestes A. Browson*, Vol. 9, AMS Press, New York (1966).

Caplan, A.: 1980, 'Have Species Become Declassé?' in *PSA 1980*, Vol. 1, P. D. Asquith and R. Giere (eds.), Philosophy of Science Association, East Lansing, MI.

Caplan, A.: 1981, 'Back to Class: A Note on the Ontology of Species', *Philosophy of Science*, forthcoming.

Cohen, I. B.: 1976, 'William Whewell and the Concept of Scientific Revolution', in *Essays in Memory of Imre Lakatos*, R. S. Cohen *et al.* (eds.), D. Reidel, Dordrecht, Holland, pp. 55–63.

Cohen, I. B.: 1977, 'History and the Philosophy of Science', in *The Structure of Scientific Theories* (2nd. ed.) F. Suppe (ed.), University of Illinois Press, Urbana, pp. 308–360.

Ghiselin, M.: 1974, 'A Radical Solution to the Species Problem', *Systematic Zoology* 23, 536–544.

Horowitz, M. C.: 1976, 'Aristotle and Woman', *Journal of the History of Biology* 9, 183–214.

Hull, D. L.: 1967, 'The Metaphysics of Evolution', *The British Journal for the History of Science* 3, 309–337.

Hull, D. L.: 1976, 'Are Species Really Individuals?' *Systematic Zoology* 25, 174–191.

Hull, D. L.: 1978, 'A Matter of Individuality', *Philosophy of Science* 45, 335–360.

Hull, D. L.: 1979, 'In Defense of Presentism', *History and Theory* 18, 1–15.

Hull, D. L.: 1980, 'Individuality and Selection', *The Annual Review of Ecology and Systematics* 11, 311–332.

Kitts, D. B. and Kitts, D. J.: 1979, 'Biological Species as Natural Kinds', *Philosophy of Science* 46, 613–622.

Morsink, J.: 1979, 'Was Aristotle's Biology Sexist?' *Journal of the History of Science* 12, 83–112.

Olby, R.: 1979, 'Mendel No Mendelian?' *History of Science* 17, 53–72.

JOHN BEATTY

REPLY TO HULL

Strange things happen to a philosopher who hangs out with historians of science. Some days you're prudishly inoffensive, apologetic, etc. with regard to the relations between the fields, some days you're defensive, and some days you're a Lakotosian reactionary. And, hopefully, some days you're just appropriately historically conscious, whatever that means.

David Hull's historiographic comments seem to me to reflect concern for my prudishness at the time. Perhaps I was unduly careful in utilizing nineteenth century views of theory-laden meaning to analyze problems of theory change in the nineteenth century. My concern for doing so, however, was that I wanted to explain certain theoretical developments in terms of my subjects' awareness of the consequences of theory-laden meaning. I did not want to rely solely on twentieth century views of theory ladenness for fear of attributing rationales to my nineteenth century subjects that they would not or could not have recognized. Nevertheless, Hull's historiographic comments still serve as a useful warning that prudishness can be pernicious, and inoffensiveness downright offensive.

Hull also raises an important matter concerning the substance of my essay. I argued that controversies concerning the evolution of species were connected to controversies concerning the reality of species. Hull points out that the evolution controversy is also linked to another controversy concerning the metaphysical status of species: the issue of whether species are individuals or kinds. As Hull and Michael Ghiselin have argued, only certain sorts of ontological entites change over time while retaining their identities: individuals do, but classes or kinds do not. Thus if, according to evolutionary theory, species change over time while retaining their identities, then, from the perspective of evolutionary theory, species must be individuals rather than kinds. That biological species are not, after all, paradigmatic natural kinds has not been generally recognized. But the problem is receiving more and more attention in the biological and philosophical literature.

While I agree with Hull that evolutionary theory occassioned a shift in the ontological status of species, from kinds to individuals, I do not see that this provides a solution to my final question, 'Were the nonevolutionists wrong?". Hull suggests that the nonevolutionists may not have been wrong

Michael Ruse (ed.), Nature Animated, 109–111.

about the evolution of species, given their theory-laden definitions of 'species', but they were wrong about the ontological status of species. But it is not clear to me why, given the nonevolutionary definition of 'species', the nonevolutionists were wrong about the nature of the referents of the term. It is not clear whether the nonevolutionists used the term to refer to individuals. But unless they used the term to refer to individuals, they would not have been wrong in characterizing the referents of the term as kinds or classes.

This issue is related to other issues that deserve some attention. The following comments are not in direct response to Hull's comments. But I would like to take the opportunity to add a sort of postscript to my essay. I now recognize that, provoked by the manner of my conclusion, one might be more inclined to consider what's not in a word. That is, convinced that the nonevolutionists were quite wrong, one might be inclined to look beyond the *words* of the evolution disputes to the *world* whose nature was disputed.

One means of proceeding in this case is first to distinguish the connotation from the denotation of a term. The connotation of a term – its definition in the usual sense – specifies the properties an entity has in virtue of which the term is properly applied to that entity. The denotation of a term, on the other hand, consists of the set of entities to which the term is applied. 'Dictophile', I recently learned, connotes a person who collects dictionaries as a hobby. The newspaper article that informed me of this also included a picture of a person denoted by that term, and suggested that there were very few other people so denoted. Similarly, we've considered at least two different connotations of the term 'species': species as breeding groups, and species as breeding groups that perpetuate their kind. If we consider connotation only, we can accept the evolutionists' claim that species evolve, without being forced to reject the nonevolutionists' claim that species do not evolve, as long as the first use of 'species' connotes breeding groups, and the second use connotes breeding groups that perpetuate their kind.

However, if we look beyond the nonevolutionists' and later evolutionists' *definitions* of 'species' to the *entities they termed* 'species', we might find that, despite the change in connotation of 'species' since the Darwinian Revolution, the denotation of the term has not changed. It seems that the most reasonable candidates to consider as the invariant denotation of the term 'species' are temporally extended breeding groups, since the breeding-group criterion is common to the two definitions of 'species' we've considered. In this case, it would have to be argued that the nonevolutionists' definition of 'species' mistakenly attributed a property to entities termed 'species' that those entities did not actually possess – namely, permanence.

But in this case, at least, evolutionists and nonevolutionists would have been talking about the same things in the world (breeding groups) and disagreeing about the nature of those things (whether or not they are permanent), so that if the evolutionists were right, the nonevolutionists were wrong.

I believe this is basically a reasonable approach to the conflict that our heart of hearts tells us occurred during the Darwinian Revolution. But as it stands, it has some problems. It takes for granted, for instance, that the nonevolutionists coupled an incorrect connotation with a correct denotation of the term 'species'. But surely a nonevolutionist could also have claimed that his denotation rather than his connotation was mistaken, if he were to find that he had termed an evolving, temporally extended breeding group a 'species'. Thus, strictly abiding by his connotation of 'species', the nonevolutionist might refuse to apply the term 'species' to many of the entities in the world that the evolutionist would call 'species'. Thus, it is not at all clear that, despite the discrepant connotations associated with the term 'species', the evolutionists' and nonevolutionists' 'species' were denotatively the same (i.e., simply temporally extended breeding groups).

On the other hand, it seems fair to say that whether or not the evolutionists and nonevolutionists fell into any straightforward conflict with regard to the evolution of *species* as temporally extended breeding groups, they nevertheless held directly conflicting theses concerning, more simply, the evolution of temporally extended breeding groups. For instance, Lyell describes the evolutionists' position in terms of the assumption of "the possibility of the indefinite modification of *individuals descending from common parents*" (1835, p. 418, my emphasis). And he describes the nonevolutionists' position in terms of the assumption that "*all the individuals propagated from one stock* have certain distinguishing characters in common, which will never vary, and which have remained the same since the creation of each species" (1835, p. 407, my emphasis).

Thus, there may have been, after all, a point of clear conflict in the Darwinian Revolution: an issue concerning which we can confidently day that if the evolutionists were right, the nonevolutionists were quite wrong. As strange as it sounds, though, the conflict was not over the evolution of *species*, but over the evolution of temporally extended breeding groups. I hope it is clear in what respects those were different issues.

Harvard University

PART III

W. R. ALBURY

THE POLITICS OF TRUTH: A SOCIAL INTERPRETATION OF SCIENTIFIC KNOWLEDGE, WITH AN APPLICATION TO THE CASE OF SOCIOBIOLOGY

1. INTRODUCTION

Are scientific ideas accepted because they are true, or are they 'true' because they are accepted? The predominant opinion within the history and philosophy of science has favored the first of these alternative views, in some cases endowing it with high moral significance.[1] In recent years, however, a somewhat piecemeal 'contextualist' or 'naturalistic' historical approach, which aims "to treat science as an aspect of our culture like any other",[2] has been gaining adherants; and the manifesto of a "strong programme for the sociology of knowledge",[3] which would account for the content of scientific knowledge in social terms, has also been published. The elaboration of these related historical and sociological positions, encouraged to some extent by recent developments in the philosophy of science,[4] has lent increasing support to the second of the two alternative views with which I opened this paper.

The nature and scope of this support has been limited, however, by the tendency of those working within the framework provided by contextualism, naturalism, or the sociology of knowledge, to concentrate almost exclusively upon relations between scientific ideas and their general social environment, leaving the more limited and immediate social environment comprised of the relevant group of scientific practitioners somewhat out of the picture. Because of this focus, the degree to which the content of scientific knowledge may be shaped by social forces has appeared highly contingent; and the suggestion has been entertained that as scientific disciplines develop and become more dominated by "technical-instrumental interests," social interests may cease to play any role at all in the formation of scientific judgements.[5]

Now such an outcome creates two major difficulties for the position in question. First of all, it means that any persuasively-demonstrated instance of the involvement of social interests in the content of scientific knowledge can easily be reinterpreted as an aberration,[6] so that the significance of historical case studies of this genre becomes extremely problematical. In addition, the way is opened for a redefinition of the sociological programme, so that it is no longer the study of the social determinants of scientific

Michael Ruse (ed.), Nature Animated, 115–129.

knowledge but rather the study of those social conditions which allow technical-instrumental rationality to produce such knowledge without undue interference.[7] These two difficulties pose a serious threat to the success of the social interpretation of scientific knowledge; however I believe they can be eliminated if more attention is given to the internal organization of science. In what follows, then, I shall attempt to outline a conception of the social determination of scientific knowledge which systematically integrates an analysis of the organization of science with a consideration of its more general social environment. The aim of this discussion will be to show that the involvement of social interests in the content of scientific knowledge is not a contingent but a necessary feature of science itself, although the particular form in which this involvement is manifested will be contingent upon historical circumstances. Finally, in order to illustrate the application of this view of science to a concrete case, a brief account of the development of sociobiology as a discipline will be sketched. This example has been chosen not only because of the biological focus of the present volume, but also because sociobiology has been the subject of a well-publicized controversy with which most readers will be familiar. Thus the distinctive features of the analysis offered here will be easier to identify by contrasting it with other positions which have been taken in the sociobiology controversy.[8]

2. THE INTERNAL AND EXTERNAL POLITICS OF SCIENCE

To avoid possible misunderstanding it is perhaps best to begin by stating that the conception to be developed here assumes the existence of a material reality which is independent of whatever humans think about it and which is both a condition of and a constraint upon human action. What is not assumed, however, is that the nature of this constraint is such that it can uniquely determine any particular account of reality; indeed, the entire thrust of the following discussion is intended to challenge any assumption of this kind.

Apart from the constraint of material reality, which functions as a purely passive, negative limitation upon scientific practice, the present conception recognizes two active, positive constraints arising from the social character of science itself to furnish the dynamic impetus for scientific development. These social constraints take the form of competitive struggles for power and resources at two levels: a struggle between individuals, disciplinary groups, etc., within science; and a struggle between the scientific community and other organized social interests within society as a whole. It will be the contention of this paper that the acceptance of scientific ideas *always* results from a compromise among these three constraints, and that in the context of contemporary science, power relations within and between scientific disciplines are *usually* the determining factors.

To begin with the constraint of material reality, let us consider the experimental situation, the most rigorous and best controlled instance of the operation of this constraint upon scientific thought. Suppose a particular theory is said to imply that a certain phenomenon should be observable under given conditions. An experiment is designed to produce these conditions and observations are made to detect the phenomenon. Now the phenomenon may or may not be observed, but in either case the effect of the experiment in either supporting or undermining the theory in question depends upon the judgement of the relevant members of the scientific community. They must judge whether the theory was correctly interpreted in the first place, whether the experimental procedures were adequate, whether the observational or measurement techniques used were sufficiently sensitive, whether the margin of error involved was within acceptable limits, and so on. Moreover, if the outcome of the experiment is judged to be inimical to the theory in question, then the decision as to which possible new theory or alteration of the old theory represents the 'best' response to the situation again depends upon the judgement of the relevant members of the scientific community. Judgements of this kind will of course be influenced by the prevailing standards within the appropriate discipline, the status of which we shall examine below. But for the moment it is sufficient to point out that the precise application of such standards to any given situation is never a clear-cut matter, and that different individuals or groups with competing interests at stake will have to interpret these standards differently if they are to remain competitive. The result of these considerations, then, is that material reality functions as a limiting factor rather than a determining factor in the acceptance of scientific ideas; and the degree to which the constraints of that reality limit scientific knowledge is determined by social processes within the scientific community which resolve conflicting interpretations of the significance of those constraints in any given case. As we shall see, these processes are largely a product of competitive struggles within the scientific community, a subject to which we now turn.

Every field of science is structured socially in such a way that there are a few acknowledged leaders whose past achievements and present command of resources give them a disproportionate influence over the judgements made in that field. The more science becomes an organized social activity, the more easily these scientific 'heavyweights' can be identified by such criteria as their selection for Nobel Prizes and other major awards, their membership in the various national and international scientific societies, their position as editors of leading journals, and other characteristics of this sort. Generally, one tends to regard these elite scientists as having achieved their position by virtue of their valuable contributions to science. But it must be remembered that the judgement which recognized their work as valuable was the judgement of the scientists who happened to constitute the elite group at the time these contributions were made; and *that* elite group got where they were because their own work was recognized by an earlier elite group. In other words, in the community of science, as in many other forms of social organization, an elite group perpetuates itself by co-opting people whom it judges to have the qualities it values most highly; and these valued qualities usually turn out to be the qualities which the elite group itself already has.

The struggle to enter this elite group, or to maintain one's competitive position within it, is vividly described by the biophysicist, Richard A. Cone, in the following terms.

> Historians and scientists often talk about science as adding bricks to the structure of knowledge, etc., as though the progress of science has to do with finding out the realities of nature and has very little to do with the sociology of science. In fact the situation is exactly the other way around. It's your adversaries, your peers, and your interaction with them, which at least for scientists tremendously regulate what you do. In fact, you are dealing with a small group of peers – a REALLY small group of peers. . . . You can have an interesting thought, but know very well that if you pursue the experiment and write it up, well, that would be nice but it would have no impact. Your work has got to be on the main line, in the area your peers have decided is a breaking new field. You've got to be on the paradigm. And once you latch onto it, it is exhilarating and powerful. Grant money comes in, publications come in, students come, invitations to talk come – if you are on the main line. But if you are trying to probe out on the edges, you are just a lonely little soul.[9]

Now in the context of this competitive struggle, the farther up the hierarchical social ladder a scientist has gone, the better chance he or she has of going higher still; because every success in the eyes of the elite group puts one in a better position to acquire more resources (in the form of grants, research assistants, experimental facilities, and so on) for the production of another success. Sociologists have referred to this phenomenon as the 'Matthew Effect' in science, citing the gospel saying: "Unto every one that hath shall be given, and he shall have abundance: but from him that hath not shall be taken away even that which he hath."[10] Ecclesiastical elites perpetuate themselves by the laying on of hands; scientific elites do it by the laying on of grants.

We should not assume, however, that the scientific elite is a homogeneous group with respect to power and influence. Being at 'the top' of one's field is defined by one's competitive position in relation to one's 'peers'; and this is why a 'conspiracy theory' of the scientific oligarchy can not succeed and why vigorous controversies are possible among the most eminent scientists. One's position in the social hierarchy of science is maintained or advanced only by continual competitive activity. Part of this activity consists in the production of research results which are judged to be acceptable by a significant proportion of the relevant scientific community: as we have seen, the more recognition and credit one's work as a scientist attracts, the higher up one advances in the scientific hierarchy. But this is only a part of the story, because the higher up one advances, the more say one has in determining what sort of scientific work deserves recognition and credit. The effect of this situation is that a scientist is able to enhance his or her own position in the hierarchy by giving recognition and credit to the work of other scientists that is most like his or her own, and by withholding recognition and credit from work that is least like his or her own. It is in this context that one can appreciate the competitive significance of such apparently 'unproductive' activities as writing review articles on the state of research in a field, writing letters of reference in support of candidates for employment or promotion, refereeing proposals for grants and articles submitted for publication, editing journals, and examining dissertations. For the scientist engaging in all these activities,

> what is at stake is in fact the power to impose the definition of science (i.e., the delimitation of the field of problems, methods and theories that may be regarded as scientific) best suited to his specific interests, i.e., the definition most likely to enable him to occupy the dominant position in full legitimacy, by attributing the highest position in the hierarchy of scientific values to the scientific capacities which he personally or institutionally possesses.[11]

The power that one has as a scientist to impose one's own definition of science upon a field varies, of course, from being practically negligible at the graduate student level to being very significant at the Nobel laureate level. But the general principle operates throughout this range – the principle that it is not only the recognition and credit that one receives that advances one in the hierarchy of science, but also the recognition and credit that one bestows upon (or 'invests in') the work of others. Hence there can be no impartial judgements in science because there is no judge who is not a party to the case. The same characteristics which qualify an individual to serve as judge in any particular case, also necessarily endow that individual with an interest in the outcome of that case. The collective judgements of the scientific community, then, can legitimately be said to result from a power struggle within that community. Scientific reputation means power within the scientific community – power not only to monopolize resources but also to influence the community's judgements and thereby to further increase one's own reputation through those judgements. It is not suggested, of course, that this power struggle occurs independently of the constraints of material reality; but what *is* suggested is that the way these constraints are investigated, and the way the results of those investigations are interpreted, are chiefly determined by the internal power struggle – the domestic politics, as it were – of the scientific community.

A final constraint which must be taken into account is that arising from the relationship between the scientific community and the rest of the society of which it is a part. Here, too, one can legitimately speak of the scientific community being engaged in a power struggle; but in this case the interests of the community as a whole are pitted against those of other elements in society. For the scientific community, the stakes involved are numerous and complex; but for schematic purposes they can be reduced to the following three categories: material resources in the form of external funding with internal autonomy as to how those funds are spent; political influence in the form of a privileged status for scientists as government advisors; and ideological predominance in the form of a virtual monopoly, within the educational system, for science as the 'correct' way to understand nature and as a basis for the ethics of individual belief. The struggle of the scientific community to maintain and advance its position on these three fronts constitutes the external politics of science. Such external politics act as a constraint upon the internal politics of science: a position within a scientific controversy which seriously compromises the external political interests of science is unlikely to succeed. On the other hand, external politics can also function as a resource in scientific controversy: a position which is seen to favor the interests of the scientific community as a whole can sometimes command assest from individual scientists whose internal political interests, narrowly conceived, would seem to dictate opposition. Such a course of events would be most likely to occur when the internal political costs of opposing the interests of the community as a whole began to outweigh the internal political benefits of defending a (personally) more advantageous position.

According to the view outlined here, then, the internal politics of science – the competitive struggle for power and influence within the scientific community – is the principal determining factor in the acceptance of scientific ideas and their certification as 'true' scientific knowledge. It should be clear from the formulation of our discussion in terms of the interaction of certain 'constraints' that the conception of science put forward here does not involve any voluntarist fantasies whereby individuals can concoct their own private worldviews and endow them with scientific status. Although in principle any view whatever can be defended, in practice the cost of defending most of them is prohibitive. It should also be clear, in spite of the occasional use of the language of intentionality, that the conception of science outlined here does not rely upon any theory of the motivations of individual scientists. The issue is not the sincerity or cynicism of particular individuals, but the types of effects which their actions have within a given social system. It is not implied that scientists are always clearly aware of their individual and collective interests, or that they always act in conformity with these interests. What *is* implied is that individual scientists who consistently act against these interests, as defined above, will be reduced to marginal status or eliminated from scientific competition altogether, and thus fail to exercise any influence upon the judgement of what is to be accepted as 'true' scientific knowledge. And conversely, individual scientists who – for whatever motives – consistently act in accordance with these interests will tend to gain increasing influence over the judgements of the scientific community as to what will be accepted as 'true' scientific knowledge. Finally, given the operation of these two contrary tendencies, it is not necessary to postulate that any individual scientist does act consistently, one way or the other; so long as it is recognized that actions of the one sort decrease the future efficacy of the individual in the scientific community, while actions of the other sort increase his or her future efficacy, then the overall patten of the development of scientific knowledge remains the same. Both the prevailing standards of judgement within science at a given time, and the content of scientific knowledge certified as 'true' in accordance with those standards, are predominantly determined by a social process of competitive struggle within the scientific community.

3. THE CASE OF SOCIOBIOLOGY

An exhaustive application of the conception developed above to the case of sociobiology would require book-length treatment, which present circumstances do not permit. For the purposes of this paper, however, it will be sufficient to give a rough sketch of how this conception applies to the development of sociobiology as a discipline and to highlight the contrast between this approach to the problem and those which have chiefly characterized the sociobiology controversy thus far. To further simplify matters, I shall not analyze the literature of this debate in any detail[12] but will treat it as embodying two 'ideal-type' positions, which I shall call the 'rationalist' position and the 'reductionist' position for ease of identification.

Sociobiology, the biological study of human and animal social behavior using the techniques of population genetics, has been the subject of an ongoing ideological and methodological critique since the publication of E. O. Wilson's major text, *Sociobiology: The New Synthesis*, in 1975.[13] The fundamental premise of sociobiology is that behavioral patterns, including those of humans, are inheritable and are selected by evolutionary pressures. This premise has led critics to charge that sociobiological theory is racist, sexist, and permeated with the values of the capitalist social system, since it represents all these characteristics of contemporary western society as biologically-determined aspects of human behavior. Despite this criticism, however, sociobiology has grown and flourished as a discipline since 1975; so the question is raised of how one can account for this disciplinary success story.

The rationalist answer to this question has stressed the intellectual merit of sociobiological theory. However questionable sociobiology may seem on particular points of doctrine, its application to human behavior is in general intellectually sound for a variety of reasons: first, because analogies can be observed between certain forms of human behavior and the behavior of other primate species; second, because the general scientific worldview since Darwin's time has included humans among the animal kingdom; and third, because bringing all forms of social behavior under a single evolutionary explanation would constitute a theoretical synthesis of the first order. From this point of view, sociobiology is already so much a part of the evolutionary 'paradigm' that working out its detailed application to unexplained aspects of human and animal social behavior is just a matter of 'puzzle-solving.'[14] The success of sociobiology as a discipline, then, depends principally upon its intellectual promise; and the fact that many of its doctrinal pronouncements run far ahead of any evidence in support of them is, according to one advocate, "probably less a weakness than it is a sign of vigor, indicating a youthful, aggressively expanding science."[15]

The alternative, reductionist answer has pointed to the ways in which sociobiological ideas serve dominant class interests in U.S. society. As the struggle of the women's movement for the passage of an equal rights amendment to the Constitution became intense, sociobiology claimed that men are innately dominant while women are innately docile and domestic. At a time when ethnic minorities pressed their demands for social and economic equality, sociobiology maintained that xenophobia is a genetically endowed component of human nature. At a time when U.S. foreign policy required a revival of military assertiveness after the relative quiescence of the post-Vietnam period, sociobiology proclaimed the "true, biological joy of warfare."[16] And the list goes on, including the basic analogy between the selfish gene and the selfish capitalist.[17] On this interpretation, then, the disciplinary success of sociobiology depends chiefly upon its legitimization of the interests of the ruling class in contemporary U.S. society.

Now the distinctive feature of the account of science developed in the present paper is that, in its application to the case of sociobiology, it incorporates the most important aspects of the rationalist and reductionist characterizations of sociobiological *theory* while rejecting the explanations offered by these two positions for the success of the sociobiological *discipline*. This feature of the present approach is nicely expressed in the following comment by Pierre Bourdieu:

> Ideologies owe their structure and their most specific functions to the social conditions of their productions and circulation, i.e. to the functions they fulfill, first for the specialists competing for the monopoly of competence in question [in this case, for scientific competence] . . . , and secondarily and incidentally for non-specialists. When we insist that ideologies are always *doubly determined*, that they owe their most specific characteristics not only to the interests of the classes or the class fractions which they express . . . but also to the specific logic of the field of production . . . , we obtain the means of escaping crude reduction of ideological products to the interests of the classes they serve (a 'short circuit' effect common in 'Marxist' critiques), without falling into the idealist illusion of treating ideological productions as self-sufficient and self-generating totalities amenable to pure, purely internal analysis.[18]

In the terms of our earlier discussion, Bourdieu's 'double determination' consists of the combined effects of the internal and external politics of science, with the internal politics (defining "the specific logic of the field of production") predominating. The reductionist account of sociobiological theory captures an important aspect of the external politics of sociobiology – namely, that its theoretical and disciplinary development is constrained by dominant class interests in U.S. society (although one need not accept that these interests have been correctly identified in reductionist critiques).[19] The rationalist account, on the other hand, captures an important aspect of the internal politics of sociobiology – namely, that the 'medium of exchange' within science consists of evidence and argumentation of an acceptable standard (although what is to count as an acceptable standard is determined by an internal political struggle). Nevertheless, the limitations of these accounts obscure the basis for the success of sociobiology as a discipline, as the following sketch will attempt to show.

3.1. *The Constraint of Material Reality*

The constraint of material reality upon a scientific discipline takes the form of empirical evidence judged to be relevant and emmeshed in a theoretical matrix of an acceptable kind. As we have already indicated, judgements of empirical relevance and theoretical adequacy can be seen as the outcome of a social struggle, with different positions representing different interests within the scientific community. On this front, the sociobiological strategy consists in taking a few relatively uncontroversial results concerning insect populations – results whose scientific acceptability has already been won in previous struggles – and generalizing them to cover all forms of social behavior in other species. The question then becomes one of defending both the acceptability of the "aggressively expanding" theoretical structure thus generated and the relevance of certain analogical evidence adduced in support of it.

Methodological criticisms of sociobiology can raise the stakes of defending it but they cannot bring this process to a halt so long as the perceived chances of winning acceptance for the descipline outweigh the risks of defending it. The more difficult the 'puzzle' is acknowledged to be, the greater the potential rewards for producing an acceptable solution to it. Thus in the defence of sociobiology, emphasis tends to be placed not upon the present record of accepted results but upon a future promise of producing results. This emphasis is clear in Wilson's claim that "since sociobiology still has a relatively weak theoretical structure, it presents the entrepreneur with unusual opportunities for discovery."[20] In this context analogical evidence serves not so much to support the theory (in the rationalist sense) as it does to support the expectations of scientific "entrepreneurs" for a profitable return on their investment.

3.2. *The Constraint of the Internal Politics of Science*

Within the scientific community the internal political struggle over sociobiology has been complicated by the fact that this "new synthesis" did not have a recognized group of specialist practitioners, so that the process of determining which members of the scientific community were most competent to judge the claims of sociobiology was itself an element in that political struggle. According to Wilson's conception, "the new sociobiology should be compounded of roughly equal parts of invertebrate zoology, vertebrate zoology and population biology".[21] The technical demands of mastering these three constituent fields – the first two characterized by their wealth of empirical detail and the third by its mathematical complexity – effectively limit the struggle for the "monopoly of scientific competence" in sociobiology to a relatively select group of practitioners, excluding therefrom ethologists and behavioral physiologists.[22] But at the same time as the access to competence in sociobiology was being narrowed, by increasing the investment required to attain it, a proselytizing campaign on behalf of the new synthesis was being carried out among such fields as anthropology, psychology and sociology[23] – fields whose practitioners would rarely succeed in attaining sociobiological competence as defined by Wilson. This effort at colonizing the human sciences can be seen, then, as a move to expand the market of consumers of sociobiology's products, rather than as an attempt to recruit actual sociobiologists.

One inducement for practitioners of the human sciences to enter into this colonial relationship with sociobiology is the promise that their competitive positions within their own disciplines will be thereby enhanced as they deploy higher-status scientific products in lower-status scientific fields. The other inducement is the cognitive equivalent of gunboat diplomacy: the threat of direct annexation. According to Wilson, "it is not too much to say that sociology and the other social sciences, as well as the humanities, are the last branches of biology waiting to be included in the Modern Synthesis. One of the functions of sociobiology, then, is to reformulate the foundations of the social sciences in a way that draws these subjects into the Modern Synthesis".[24] The implication, of course, is that if practitioners of these sciences are not to become altogether obsolete, then their only recourse is to become consumers of sociobiological products.

Thus far we have identified three strong incentives for the support of or conversion to sociobiology by specialist practitioners of the biological sciences: (1) the definition of a unique product, sociobiological knowledge, and the exhibition of a prototype model of its production in relation to insect populations, together with the promise of expanded production in the future; (2) an indication that the production of this product can be subjected to intense monopolization; and (3) the creation of a market for this product among other fields, assuring the scientific legitimacy of the producers' monopoly by showing that their product has scientific application outside the charmed circle of its origin. These three inducements for the adoption and defence of sociobiology by those capable of becoming recognized as competent in this field, are reinforced by the role which the discipline is capable of playing in the external politics of science, to which we now turn.

3.3. *The Constraint of the External Politics of Science*

Advocates of sociobiology, writing in both the scholarly literature and the mass media, have not been reticent about publicizing the field's potential value "for the planning of future societies".[25] That such claims about policy implications are consonant with the political demands made upon scientific research is demonstrated by the guidelines which the N.S.F. distributes to referees of proposals submitted to it for research grants. Among the selection criteria which referees are to consider in evaluating proposals is the degree to which the research project under review might "assist in solving societal problems".[26] In this context it is clear that the policy claims made about sociobiology not only give it a competitive advantage in the struggle for material resources within the scientific community, but also enhance that community's competitive position within society generally, by promising a scientific product of direct utility to social planners.

It is worthy of note, however, that within the U.S. political system "the reliance on scientifically 'certified facts' has been a matter of determining not merely the content of decisions but also their public credibility and legitimacy".[27] In the light of this dual political function of American science, the unprecedented campaign in the mass media on behalf of sociobiology can be seen as a method of creating public acceptance of sociobiologists as government advisors. It is at this level, I would suggest, that the conservative aspects of sociobiological theory come into play as a reassurance that the familiar social order is not under threat from the new discipline. Similarly, a public demand for – or at least acquiescence in – the introduction of sociobiology into the educational curriculum is fostered by the conservatism of certain sociobiological doctrines.

The claimed "technical-instrumental" utility of sociobiology for social engineering purposes, and the focus on conservative theoretical elements in sociobiological popularizations, create a system of external incentives for the social support of this newly-emerged discipline. Such external support raises the career costs of opposing the development of sociobiology as a discipline, without correspondingly increasing the career benefits of such opposition. These external circumstances, then, serve to enhance the competitive position of sociobiology within the internal politics of science by identifying the disciplinary interests of sociobiology with the material, political and ideological interests of the scientific community as a whole. Nevertheless, they cannot in themselves confer a scientific victory upon that discipline; for, as Bourdieu has noted, the victory of any scientific faction will not be recognized by other scientists as a victory for *science* unless it is won according to the internal rules of the scientific community.[28] And these rules, as we have already seen, are the result of the internal politics of science. Thus it is neither the intellectual merit of its theory nor its external political utility that accounts for the success of sociobiology as a discipline, but rather its competitive advantage in the internal political struggle within science itself – an advantage which we have outlined above as a system of entrepreneurial incentives inducing scientists to support the development of sociobiology and to endow the knowledge it produces with the status of scientific truth.

University of New South Wales

NOTES

[1] See, for example, Sheffler, 1967, pp. 4–8.
[2] Barnes and Shapin, 1979, p. 9.
[3] Bloor, 1976, pp. 4–5.
[4] See, for example, Mulkay, 1979, ch. 2.
[5] Shapin, 1979, p. 65.
[6] Shapin (*ibid.*) explicitly recognizes this possibility in connection with his own studies of phrenology.
[7] See, for example, Meynell, 1977.
[8] The conceptions developed in this paper owe a great deal to published works by Michael Mulkay and Pierre Bourdieu, and to unpublished papers by Camille Limoges and Colin Gunn. A joint study by Gunn and myself, currently in preparation, will elaborate on the interpretation of sociobiology outlined here.
[9] Cone *et al.*, 1980, pp. 23–24.
[10] Merton, 1973, p. 475. See also Mulkay, 1976, p. 449.
[11] Bourdieu, 1975, p. 23.
[12] For a review of this literature see Albury, 1980.

[13] See also the later abridgement of this text: Wilson, 1980.
[14] The use of Kuhnian terminology at this point does not, of course, presuppose general acceptance of Kuhn's account of the nature of scientific development (Kuhn, 1970).
[15] Barash, 1978, p. 19.
[16] Wilson, 1975, p. 573; Wilson, 1980, p. 298.
[17] Anonymous, 1978.
[18] Bourdieu, 1977, p. 4.
[19] In particular, the reductionist emphasis on sociobiology's theoretical legitimization of the social *status quo* seems to miss the important role which the discipline is designed to play at the technical level as a tool for social engineering. For a discussion of this aspect of sociobiology, to which we shall return below, see Haraway, 1979.
[20] Wilson, 1975a, p. 5.
[21] Wilson, 1975, p. 4; Wilson, 1980, p. 4.
[22] *Ibid.*
[23] Alper *et al.*, 1978, p. 478.
[24] Wilson, 1975, p. 4; Wilson, 1980, p. 4.
[25] Wilson, 1975, p. 548; Wilson, 1980, p. 272. See also Wilson, 1979, p. 99.
[26] National Science Foundation, 1979.
[27] Ezrahi, 1971, pp. 121–122.
[28] Bourdieu, 1975, p. 21.

REFERENCES

Albury, W. R.: 1980, 'Politics and rhetoric in the sociobiology debate', *Social Studies of Science* 10, 519–536.

Alper, J., Beckwith, J., and Miller, L. G.: 1978, 'Sociobiology is a political issue', in Caplan, 1978, pp. 476–488.

Anonymous: 1978, 'A genetic defense of the free market', *Business Week* (April 10), 100, 104.

Barash, D. P.: 1978, 'Evolution as a paradigm for behavior', in Gregory *et al.*, 1978, pp. 13–32.

Barnes, B. and Shapin, S. (eds.): 1979, *Natural Order: Historical Studies of Scientific Culture*, Sage, Beverly Hills/London.

Bloor, D.: 1976, *Knowledge and Social Imagery*, Routledge and Kegan Paul, London.

Bourdieu, P.: 1975, 'The specificity of the scientific field and the social conditions of the process of reason', *Social Science Information* 14 (6), 19–47.

Bourdieu, P.: 1977, 'Symbolic power', in *Two Bourdieu Texts*, trans. R. Nice, University of Birmingham Centre for Contemporary Cultural Studies, Birmingham.

Caplan, A. L. (ed.): 1978, *The Sociobiology Debate: Readings on the Ethical and Scientific Issues Concerning Sociobiology*, Harper and Row, New York.

Cone, R. A., Huggins, W., Macksey, R., Pond, R., Shore, D., and Hancock, E.: 1980, 'Scientific Creativity', *Johns Hopkins Magazine* 31 (1), 20–25.

Ezrahi, Y.: 1971, 'The political resources of American science', *Science Studies* 1, 117–133.

Gregory, M. S., Silvers, A., and Sutch, D. (eds.): 1978, *Sociobiology and Human Nature: An Interdisciplinary Critique and Defense*, Jossey-Bass, San Francisco.

Haraway, D.: 1979, 'The biological enterprise: sex, mind, and profit from human engineering to sociobiology', *Radical History Review* 20, 206–237.

Kuhn, T. S.: 1970, *The Structure of Scientific Revolutions*, 2nd ed., University of Chicago Press, Chicago.

Merton, R. K.: 1973, *The Sociology of Science*, University of Chicago Press, Chicago.

Meynell, H.: 1977, 'On the limits of the sociology of knowledge', *Social Studies of Science* 7, 489–500.

Mulkay, M. J.: 1976, 'The mediating role of the scientific elite', *Social Studies of Science* 6, 445–470.

Mulkay, M. J.: 1979, *Science and the Sociology of Knowledge*, Allen and Unwin, London.

National Science Foundation: 1979, *NSF Criteria for the Selection of Research Projects* (F. L. 100, 2–79), [Government Printing Office,] Washington.

Shapin, S.: 1979, 'Homo phrenologicus: anthropological perspectives on an historical problem', in Barnes and Shapin, 1979, pp. 41–71.

Sheffler, I.: 1967, *Science and Subjectivity*, Bobbs-Merrill, Indianapolis.

Wilson, E. O.: 1975, *Sociobiology: The New Synthesis*, Harvard University Press, Cambridge, Mass.

Wilson, E. O.: 1975a, 'Some central problems of sociobiology', *Social Science Information* 14 (6), 5–18.

Wilson, E. O.: 1979, [Interview,] *OMNI* (February), 96–99, 134–136.

Wilson, E. O.: 1980, *Sociobiology: The Abridged Edition*, Harvard University Press, Cambridge, Mass.

PART IV

M. EAGLE

ANATOMY OF THE SELF IN PSYCHOANALYTIC THEORY

INTRODUCTION

Certain psychoanalytic formulations, because they represent challenges to traditional modes of thought, have been of interest to philosophers. The psychoanalytic idea of unconscious mentation, for example, represents a challenge to the traditional equation of mental and conscious. For someone making such an equation, the notion of unconscious mentation would seem absurd, a contradiction in terms. And indeed, quite a number of philosophers found talk of unconscious beliefs, desires, motives, and wishes, at best, somewhat odd, and at worst, absurd and nonsensical (to cite just a few, Field *et al.*, 1922; Siegler, 1967; Goldman, 1970).

The related question of the psychoanalytic partitioning of the personality has also been of interest to philosophers, in particular in relation to the unity of self issue (e.g., Fingarette, 1969; Moore, 1979; Sartre, 1956; Thalberg, 1976). As Perry (1975) has observed, since the classic expositions of Locke (1689) and Hume (1739), most philosophers concerned with unity of self have addressed the issue of *identity over time* rather than *identity 'at a time'*. Only recently, have they become interested in the latter. By contrast, psychoanalysis takes identity over time for granted and raises questions about the nature of identity 'at a time'. A well-known feature of psychoanalytic theory is its anatomizing of the personality – first topographically in terms of unconscious, preconscious, and conscious and later structurally, in terms of id, ego, and superego. This anatomizing of the self has been held by some to violate the traditional and commonsense idea of the unity of self and commonly accepted ideas regarding the nature of a person. Thalberg (1976), for example in a recent paper concludes that such anatomizing will not work and "that we seem to have no alternative but to assume the unity of the self" (p. 171). Later in the paper, I will examine Thalberg's arguments, but it might be useful to remind ourselves of the historical context in which the psychoanalytic anatomizing of the self developed and of the kind of clinical phenomena which appeared to necessitate such a radical conception of personality.

Michael Ruse (ed.), Nature Animated, 133–161.

HISTORICAL BACKGROUND

While the particular rendering may have been distinctively psychoanalytic, the general idea of the partitioning of the self was not unique to psychoanalysis, but had its clear roots in the attempts of late 19th-century dynamic psychiatry to explain such phenomena as hypnotic manifestations, somnambulism, automatic actions, multiple personalities, fugue states, and hysterical conditions. After observing these various dissociative phenomena, the central conclusion of Charcot, Janet, Binet, and others was the basic idea that "split-off fragments of the personality could follow an invisible development of their own and manifest themselves through clinical disturbances" (Ellenberger, 1970).

As Ellenberger (1970) notes, all the above phenomena were believed to represent means of gaining access to the hidden, unconscious mind. In 1890, Dessoir wrote a book called *The Double Ego* in which he argued for the existence of two egos, an upper consciousness (überbewusstein) and an under consciousness (unterbewusstein). In 1868, Durand (de Gros) proposed the doctrine that the human organism "consists of anatomical segments, each of which had a psychic ego of its own, and all of them subjected to a general ego, the ego in chief, which was our usual consciousness" (Ellenberger, 1970, p. 146). (Note the congruence between this formulation and recent conclusions based on split-brain patients – see Nagel, 1971).

In hysteria, the fragments split off from the personality or, as Charcot referred to them, "fixed ideas" are created by two facts: one, the trauma (to which the hysterical symptoms could presumably always be traced) created a hypnoid state analogous to that produced in hypnosis; and two, the patient's constitutional predisposition to hypnoid states or "narrowed field of consciousness." Ideas experienced in this dissociated hypnoid state are likely to develop autonomously, isolated from the rest of the personality, and can then wreak the kind of havoc on the rest of the personality as is observed in hysterical symptoms and other dissociative phenomena. In other words, as Ellenberger (1970) states, in describing Janet's formulation, "subconscious fixed ideas are both the result of mental weakness and a source of further and worse mental weakness' (p. 367).

It will note noted that the model for the "narrowed field of consciousness" to which the hysterical patient is chronically predisposed is the hypnotic state and that the model for "unconscious fixed ideas" is the hypnotic suggestion. Indeed, Charcot described the latter in terms indistinguishable from the "fixed ideas" primary in hysteria. According to Charcot, through hypnotic suggestion, "an idea, a coherent group of associative ideas settle themselves in the mind in the fashion of parasites, remaining isolated from the rest of the mind and expressing themselves outwardly through corresponding motor phenomena. . . . The group of suggested ideas finds itself isolated and cut off from control of that large collection of personal ideas accumulated and organized from a long time, which constitutes consciousness proper, that is, the Ego" (Ellenberger, 1970, p. 149). Whatever the cause of the "fixed ideas," their isolation from the rest of the personality and their autonomous development are the proximate and decisive factors in hysteria and other illnesses. In very strong terms, Janet (1889) tells us that "one should go through the entire field of mental diseases and a part of physical diseases to show the mental and bodily disturbances resulting from the banishment of a thought from personal consciousness" (p. 436).

All these formuations and developments preceded psychoanalytic theory. Indeed, Freud came upon an intellectual scene which was somewhat dominated by the notions that the personality can be divided and that split-off fragments of the personality, such as an idea implanted by hypnosis or ideas and affects associated with traumas, are endowed with an autonomous life and development and can be manifested in clinical disturbances. What came to be distinctively psychoanalytic was the formulation of the *dynamic* unconscious – that is, the beliefs that the split-off ideas and affects are actively or purposively extruded from consciousness and that they are especially linked to infantile sexuality. As the following passage from some of his earliest writings shows, Freud (1893–1895) adds to the then prevalent notions of "fixed ideas' and "splitting of consciousness" the active and intentional role of the ego: "The actual traumatic moment, then, is one at which the incompatibility forces itself upon the ego and at which the latter decides on the repudiation of the incompatible idea. That idea is not annihilated by a repudiation of this kind, but merely repressed into the unconscious. When this process occurs for the first time there comes into being a nucleus and center of crystallization for the formation of a psychical group divorced from the ego – a group around which everything which would imply an acceptance of the incompatible idea subsequently collects. This splitting of consciousness in these cases of acquired hysteria is accordingly a deliberate and intentional one" (p. 123).

This more active and purposive role given to the ego complicated and deepened the conceptual challenge to ideas of unity of the personality. One can say, as Charcot and Janet did, that whatever renders an individual more susceptible to hypnoid states (including predisposing constitutional defects) will make more likely the development of split-off ideas, which then continue their dissociated and autonomous development and are manifested in various symptoms. This more conservative position was essentially the one taken by Breuer (1893–1895) who argued that it was the experience of trauma while in a hypnoid state that mainly accounted for the development of hysterical symptoms. While partially going along with Breuer in the earliest of his writings, Freud soon rejected the idea of hypnoid states and insisted that the hysterical patient was actively extruding from consciousness ideas which were morally repugnant and generally unacceptable to the conscious ego. In other words, the reason that certain ideas were split off was not because they were experienced when the person was already in a dissociated, hypnoid state. Rather, the central idea proposed by Freud was that out of personal motives these ideas were *rendered* dissociated.[1] In short, the reasons for the splitting off of certain ideas was to be sought, not in constitutional weakness or peculiar hypnoid states of consciousness, but in the interplay of wishes, desires, and aims on the one hand and guilt, shame, and moral standards on the other. Strange hysterical symptoms could now be seen as quasi actions in which an individual seeks particular goals, reaches compromises of various sorts, and expresses certain meanings. That is, these symtoms could be understood in terms of ordinary reasons and motives, as one understands any other kind of behavior.

I say that this formulation complicates and deepens the challenge to concepts of unity of the self because in this account all aspects of the complex story are purposive in nature and are occurring simultaneously within the same person. According to the Freudian view, occurring simultaneously are purposive mental maneuvers we normally believe can only be consciously carried out by a person – including strivings for gratification, avoidance of displeasure, and reaching of compromises – which are now claimed to be carried out unconsciously and are attributed to partial components constituent of the person. Furthermore, while in prepsychoanalytic thinking, the importance of dissociated ideas was discussed almost exclusively in relation to individuals particularly susceptible to hypnoid states, in psychoanalytic formulations, the conflict between unconscious sexual wishes (and later, aggressive ones) and internalized moral standards came to be considered the universal human condition. Indeed, the very definition of personality in terms of constituent parts – id, ego, and superego reflected and, so to speak, enshrined the universal, inevitable nature of such conflict. It is this general anatomizing scheme that is the object of recent criticism by Thalberg (1976) and others (e.g., Moore, 1979; de Sousa, 1979).

THALBERG'S CRITICISMS

I come now to an evaluation of the psychoanalytic anatomy of the self. As noted, I will take as a starting point Thalberg's (1976) recent thoroughgoing criticism of Freudian partitioning schemes. Thalberg raises the basic challenge to any anatomizing proposal with his conclusion that "whatever anatomy of the self you propose, I believe you will be saddled with doctrines which are either not germane to the psychological phenomena, or else not cogent. Those in search of edification might conclude that we seem to have no alternative but to assume the unity of the self" (p. 171). A close examination of Thalberg's arguments will reveal that they do not warrant the conclusion that we "have no alternative but to assume the unity of the self." Rather, they demonstrate the difficulties with partitioning schemes which blend talk of mechanisms and of conscious agency and personify hypothesized structures of the mind. Thus, Thalberg correctly objects to explanatory schemes in which one talks, for example, about instincts striving for gratification and the ego's wish to sleep. As Thalberg and, more recently, Moore (1979) have pointed out, such talk leads us into meaningless questions such as those concerned with who, "among the denizens of our mind," is a responsible agent. One wants to counter this kind of talk by pointing out that persons strive and wish, not hypothetical structures (see Schafer, 1976). But, as far as I can see, this kind of argument pertains to the particular personifying characteristics of the Freudian partitioning scheme rather than to the general issue of unity vs disunity of the personality.

Thalberg has not told us why we might have no alternative but to assume the unity of the self nor does he ever tell us precisely what he means by unity of the self. An examination of his arguments suggests that what he intends by unity of self is *numerical* unity – one self versus multiple selves. That is, he has tried to demonstrate that schemes characterized by personified structures, homunculi, and multiple miniselves do not work and that we are best off returning to the traditional assumption of one self per customer. Or, as Moore (1979) puts it, one person per body (see Williams, 1973). But this sensible scheme of of one self or one person per body does not address itself to dimensions of unity (and disunity) of self and personality which are of special interest to psychoanalytic theory.

What needs to be said first is that implicit in the partitioning schemes of psychoanalytic theory (as well as, we shall see later, of other theories of personality) is a critical distinction between the narrower concept of self and the broader concept of personality. Although following the usage of others, I have been talking about anatomizing of the self, it is anatomizing of the *personality* that is involved in psychoanalytic theory. That is to say, notwithstanding the assumption that each of us is best thought of as a single self and single person, psychoanalytic theory directs our attention to those unconscious and disavowed activities which are not directed by a conscious agency, which are experienced as external to the self, but which nevertheless represent intelligible and intentional aspects of one's personality. In other words, while the concept of self refers mainly to these activities and strivings which are acknowledged and endorsed, the concept of personality is meant to embrace the full range of our intentional and intelligible activities, whether avowed or disavowed, endorsed or dissociated. It is the *non-equivalence* of self and personality which has led, in the first place, to discussions of unity of self. Partitioning schemes such as id, ego and superego direct our attention to the existence of disavowed and unconscious psychic trends (and functions) within the personality which exist outside the subjective self and are disavowed by one's conscious self-organization. *This* is the essential point implicit in the Freudian structural division of personality between ego and id, between *Das Ich* and *Das Es*. It can be taken to reflect the fact that certain psychic trends are experienced as not-I, as impersonal 'it.' And it is the presence of these impersonalized and disavowed psychic trends – that is, the dissociation between self and personality – which, both for pre-psychoanalytic and psychoanalytic thinking, constitutes the challenge to any simple assumption of the unity of personality.

Once one accepts the above distinction between self and personality, it becomes apparent that posing the question in terms of unity of self is somewhat misleading. For it is not unity of disunity of self, but unity or disunity of *personality* with which traditional psychoanalytic theory has been primarily concerned.[2] The conceptual and philosophical problem, in the form of the unity of self issue, enters insofar as it has been tempting to suppose that just as a person or self is 'behind' the desires, aims, rules, etc., consciously experienced, so it must follow that minipersons or other selves must be pursuing aims, following rules, etc., not available to consciousness. Or to put it in a somewhat different way, the implicit assumption is made to the effect that mental events, like pursuing aims and having desires, are the kinds of things that only conscious minds can do and that only can be done via consciousness. For example, in discussing split-brain phenomena, Nagel (1971) is led to question the unity of consciousness because "what the right hemisphere can do on its own is too elaborate, too intentionally directed and too psychologically intelligible to be regarded merely as a collection of unconscious automatic responses" (p. 235). (See also Puccetti, 1973.) That is, mental events that are complex, intentional, and psychologically intelligible are only possible when 'directed by' a conscious mind. Since, however, the mind associated with the conscious self[3] is unaware of and even disavows these complex, intentional, and intelligible mental events, there must be an additional mind 'directing' these events. Whether one is talking about split-brain phenomena, the unconscious mental processes of psychoanalytic theory, or the dissociative phenomena of 19th-century dynamic psychiatry, the above, I believe, is an essential aspect of the reasoning underlying suggestions of multiple centers of consciousness, multiple minds, and multiple selves.

This reasoning is most clearly stated in regard to split-brain phenomena. Put very simply, it is as follows: Since the right hemisphere can comprehend, carry out intentions, etc. it too (along with the left hemisphere) must be a person because only persons can carry out these sorts of things. (A variant of this argument is that it must be conscious since only conscious minds can carry out these sorts of things.) The analogous reasoning in other areas is that since one can unconsciously comprehend and engage in all sorts of intentional activities (some of which one consciously disavows), there must be additional selves or minds responsible for those activities.

I believe that the essential point to be made in response to Nagel's and others' concern is that complex, intelligible and psychologically intelligible mental processes occur outside of awareness and do not require the 'direction' of a conscious self. The assumption that only persons or mini-selves or mini-minds can carry out elaborate 'intentionally directed' and 'psychologically intelligible' activities may be confused and mistaken. Stated very simply, it may be that *parts of a person* (in more contemporary language, subsidiary control centers not requiring consciousness) can do these sorts of things. While it may be true that the very discussion of elaborate intentionally directed and psychologically intelligible activities makes sense only in the context of a person (and the animal equivalent of person), it does not follow that every time these things are done one must posit, so to speak, a complete person doing them. What this lead to is the need to generate other persons (or minds or selves) when the person I say I am denies or is not aware of doing these things. Another way of stating this is to say that while the proper context for talking about intentionality, intelligibility, etc. may be a person, it does not follow that whenever one comes across such activities there is a complete mini-person directing them. Rather, these activities and their associated control centers are part of the person. It is the *organization and integration* of all these events and processes that constitute the person. If I show behavior which is contrary to and disavowed by what I can consciously experience and report, it does not mean that one must posit a mini-person to account for this behavior. It means that the kind of person I am does not entail the fully successful and complete integration of all my intelligible and intentional activities.

It seems to me that the positing of multiple centers of consciousness, multiple minds, and multiple selves to account for complex intelligible and intentional activities of which one is unaware and which one may disavow is logically linked to the traditional philosophical equation between the mental and the conscious. That is, if mental and conscious are equated, then an *additional* center of consciousness (or mind or self) must be invoked to account for those mental activities of which one is unaware and which one disavows. This need disappears once one overcomes the traditional equation of the mental and the conscious and accepts the disjunction between the two. There is ample evidence warranting such a step. Outside the psychoanalytic and split-brain contexts – that is apart from phenomena of unintegrated and disavowed cognitions and intentions – there is a large body of data which indicates that complex, intelligent, even highly elegant responses suggesting logical, problem-solving maneuvers occur regularly outside conscious awareness. That is to say, just as how I define myself does not exhaust who I am as defined by the aims, desires, etc. revealed in my behavior so too what I report and say I know and perceive does not exhaust what, it can be shown, I have processed and discriminated.

Consider first certain perceptual experiences: Rock (1970) and others (e.g., Gregory, 1971) have presented convincing evidence that the best way to understand certain perceptual phenomena is as the product of cognitive events of the nature of hypotheses, "taking-into-account" processes, which "intervene between reception of the proximal stimulus and the resulting perception" (Rock, p. 8). These processes are, in form, very much like logical inferences although the perceiving person does not, of course, make conscious inferences.

Let me describe a concrete example. The principle that is most predictive of when one will experience induced or stroboscopic movement is: "movement will be perceived whenever the total information available adds up to the inference that an object has changed its location (Rock, 1970, p. 9). Thus, if a triangle (or any other object) is flashed on at point A in a visual field and then, a short time later at point B, one will experience the triangle moving from A to B, even though no actual movement has occurred. In other words, the implicit reasoning is elegantly simple: If an object is now here in the field and then, a moment later, there, it must have moved. Further, one can eliminate the experience of induced movement by, so to speak, "forcing" a different inference. Thus, if simultaneous with the flashing on of B, A also reappears in its original location, one will experience no movement. In short, the spontaneous perceptual experience will follow the inference that best fits the available experience. It will be identical to what would occur if one engaged in a conscious, problem-solving inference that represented the most elegant solution to a puzzle. As Rock (1970) concludes, "perception turns out to be shot through with intelligence" (p. 10). And yet, of course, no conscious inference occurs, a consideration which led Helmholtz (1962) to suggest that "unconscious inference" was operative in phenomena such as those described above. For many years, the concept of 'unconscious inference' was thrown into disrepute because it was believed that inferences could, of course, not be unconscious (here is another example of the equation between the mental and the conscious). But the assumptions that such inferences do occur best fits a wide range of perceptual phenomena. The point in the present context is that quite apart from psychoanalytic concerns and quite apart from dissociative phenomena, complex, intentionally directed and highly intelligible processes occur as a matter of course outside awareness.

As another simple example of unconscious cognitive processes, I have been struck recently by the mundane but indisputable evidence that one can have a specific expectation to which one has no conscious experiential access. Many people have had the experience of getting on an escalator which was not running and which they know is not running. Yet, they experience a kind of stumbling vertigo which is clearly a function of the discrepancy between their expectation (that the escalator would move) and the non-moving escalator. Further, one can continue to have the vertigo experience in moderated form, a number of times, although each time one knows the escalator is not running. Now, if one asks oneself, prior to the first experience, to list all the expectations of the world one has, one would certainly not have included the expectation that metal stairs, constructed in such and such a way and showing such and such features, move; and yet, one's experience and behavior indicate clearly that one does have such a specific expectation. And I do not believe that one can escape this conclusion by reverting to physiology or to talk about automatic conditioned responses. For, among other things, it is clear that before one's expectation can become apparent (in the experience of vertigo), one must have processed the objective stimulus situation to yield the highly meaningful percept: 'Escalator'.

For some final examples, I borrow from Robinson (1976), who cites them to demonstrate that simultaneous awareness and non-awareness is not unique to split-brain patients, but is rather a common occurrence.

If an observer is presented with a brief flash closely followed by one of greater intensity and duration, he will fail to report the first; this despite the fact that the first is unfailingly reported when no second flash is delivered. The effect is called backward masking and it has played an important part in modern theories of visual perception (Raab, 1963; Robinson, 1971). If, under these backward masking conditions the observer is instructed to press a key as soon as he sees a flash, we can determine the effect the mask has on the processing of the first flash. As it happens, reaction-time to the first flash is unaffected by a presentation of that very masking flash which eliminates verbal evidence of detection (Fehrer and Raab, 1962). We might say that the observer has the ability to respond as quickly to a flash he fails to 'see' as to one he does. That is, although he is not aware of the flash he can and does respond to it all the same . . .

In another class of experiments, a matrix of letters is briefly presented to an observer. The matrix may, for example, be 8 × 4 or 4 × 4. With very brief exposures it is found that the normal observer correctly reports between five and nine of the sixteen (or more) items. Loosely, we may say that the span of perception-with-retention is, ceteris paribus, about seven items wide. On any given presentation, there will be some ten items apparently unperceived. Suppose, however, that immediately after the display disappears, a cueing ring is projected over one of the matrix locations. We now find that the observer can tell us the letter that previously occupied this location even though, on the trial in which the letter occurred, he failed to report it (Sperling, 1963). Here again the observer simultaneously 'knows' and does not 'know' that the letter in question was part of the array. Without the cue, he will swear that he has reported everything he has seen. With the cue his span of retention is widened, ex post facto. (p. 74)

In addition to the evidence suggesting that we make inferences, have expectations, and process and discriminate information without awareness, there are all those examples in which one carries out purposive, complex activities (such as driving a car for long distances) with a low level of introspective awareness. In these situations, as Armstrong (1973) notes "one must in *some sense* have been perceiving, and acting purposively . . . But one was not conscious of one's perceptions and one's purposes" (pp. 93–94).

The main point to be made by all these examples is that complex, intentional, and intelligible processes can occur with either minimal or no conscious awareness. They remind us that cognition and mentation are ubiquitous and, as Freud (1900) noted, that only a portion of these ubiquitous mental events are consciously experienced. Once one accepts this disjunction between the mental and the conscious as a given, one does not need to posit a quasi-conscious mini-self or homunculus responsible for these highly intelligent processes of which the person himself is unconscious. Rather, complexity, intentionality, and intelligibility are natural and integral aspects of our engagements in the world and do not require the direction or intervention of a conscious self. Indeed, one could argue that the latter is an exceptional rather than a necessary component of our engagements. That is, the engagements which are part of one's conscious self-organization constitute only a small part of our total intentional and intelligible engagements in the world.

In the above examples, the unconscious processes involved are not in conflict with the conscious self, but are an integral part of normal adaptive functioning. Also, the mental events involved in some of the examples are, by their nature, unavailable to conscious experience. Thus, the most careful introspection and the most attentive effort will not yield a phenomenal experience of those inferential processes leading to the perception of stroboscopic movement or the vertigo experience on the stationary escalator. In the psychoanalytic context, however, the assumption is made that much of what is unconscious – certain wishes, impulses, and aims – could be consciously experienced were it not for various defensive processes which, in one fashion or another, involve the *disavowal* of unconscious wishes and aims. It is disavowal and its consequences that are of distinctive interest in the psychoanalytic context and are especially relevant to considerations of unity of personality.

AVOWAL-DISAVOWAL

Of the wide range of strivings, desires, and aims we pursue, meanings we decipher, rules and values we follow, and valuations we make – all of which are complex, intentionally directed and psychologically intelligible activities revealed in our behavior – only a part are, to use Fingarette's (1969) phrase, 'spelled our', endorsed, and made part of the self. Or, as was stated above, it is a natural state of affairs that on a pre-reflexive level, we are intentionally engaged in the world (see Sartre, 1956). Only a portion of those engagements – those 'spelled out', avowed and endorsed – come to constitute the conscious self. Other aims, meanings and values may be contrary to the constituted conscious self. In order to preserve unity and integrity of self, we dissociate and disavow aims, etc., radically at variance with one's self-organization – that is, with that structure of aims, goals, values, beliefs, etc., that one experiences as oneself (Klein, 1976). But insofar as the disavowed wishes and aims continue to influence behavior, these attempts to preserve self-unity through dissociation do not entirely work. One side of the conflict lies buried and unsusceptible to integration within the rest of the personality.

Most important, however, when a conflict is 'resolved' by disavowing aspects of it, one has rejected certain of one's strivings and aims as "not me," as external to oneself (see Frankfurt, 1976). One has dissociated aspects of one's personality from one's self-organization. *This* is the decisive aspect which threatens unity to personality, not simply the presence of conflict or whether there is awareness or unawareness (particularly, since such dissociation can be implemented with some conscious wishes, thoughts, aims, etc.).[4] A person beset by conflicting strivings and wishes, all of which he avows, is perhaps a troubled person, in distress, and in one important sense, unintegrated. But all the conflicting elements remain *personal* elements, belonging to the same person. It is the disavowal, the impersonalizing of what are, in the larger sense, one's own strivings and wishes that is the core of the threat to unity of the personality.

As Fingarette (1969) points out, in considering issues such as self-deception and defense, exclusive emphasis has been placed on knowing versus not knowing, awareness versus unawareness. Fingarette suggests, correctly I believe, that at least an equally important vantage point from which to view the issue of unconscious, defended against wishes and aims would use the language of spelling our versus not spelling out, avowal versus nonavowal, and acknowledgement versus nonacknowledgement. What is central here is that certain engagements are made explicit and acknowledged as part of the self, while others are not spelled out and not acknowledged as part of the self.

This is essentially the point of view expressed by Klein (1976), in his recent revamping of psychoanalytic theory. In this reinterpretation of defense, Klein is more fully developing Freud's (1940) latest ideas, which death prevented him from elaborating. In his very last paper, never competed, Freud discusses defense, not in terms of awareness versus unawareness, but in terms of splitting of the ego and avowal versus disavowal (see Fingarette, 1969 for a further discussion of this issue). Even prior to that, one could detect a move in this direction in the shift from "making the unconscious conscious" to "where id was, there shall ego be" as the primary goal of psychoanalytic treatment. Particularly if one interprets 'id' and 'ego' closer to the original German as the 'it' (Das Es) and the 'I' (Das Ich), the latter version of the psychoanalytic goal can be taken to mean that one must render what was originally experienced as *nonpersonal*, as an 'it', into the *personal* experience of an 'I.' Becoming consciously aware may be a necessary ingredient in this process, but it is not a sufficient one. One also needs to avow, endorse, and integrate as part of the self what was formally alien, impersonal, and to borrow Frankfurt's (1976) term, external to the self. This, I believe, is the central insight in the shift to the goal of "where the id was, there shall ego be".[5]

What is implied in the above is that conscious and unconscious conflicts can be equally threatening to unity of the personality. For what ultimately threatens unity of personality is not conflict *per se*, but the disavowal and dissociation of certain psychic elements in the conflict so that they, so to speak, stand outside self-organization and remain unintegrated. (It will be recalled that to Charcot and Janet it was the isolation of 'fixed ideas' from personal consciousness that threatened self-integrity). That one cannot have A and not A simultaneously is a fact of the world, but in no way does it suggest that one cannot want A and not A simultaneously (what is mainly involved here are conceptions of a rational agent rather than of unity of self). Clearly, a single person can want A and not A simultaneously and not present any special conceptual problems regarding unity of self. Challenges to unity of personality enter the picture when either A or not A are disavowed, dissociated and function intelligibly and purposively outside one's avowed self-organization.

That disavowal, rather than the presence of conflict or of unconscious aims, is the critical threat to unity of personality, can be seen dramatically in the case of multiple personalities. For certain people, disavowed ideas, aims, etc., become so coalesced and organized and play such a forceful role in behavior that it is as if there were an alter self co-existing with the normal conscious self. If one defines self in terms of organization of aims, values, etc., then the evidence in behavior of an organized set of disavowed aims, values, etc., different from and contrary to one's dominant self-organization, would tempt one to speak of an alter self (see Fingarette, 1969). Cases of multiple personalities are the clearest expression of this phenomenon. The style and content of behavior that is manifest at one time is so contrary to the style and content of the consciously avowed 'normal' self that one appears forced to conclude that side by side with this 'normal' self there exists an entire alternate self.

In the case of multiple personalities, the particular degree of lack of integration and kind of lack of integration (that is, the coalescing of dissociated ideas, trends, etc., to form an alternate coherent organization) is so extreme it challenges the very assumption of numerical unity of self. As the term multiple personality suggests, there appear to be at least two sets of self-organizations (that is, two coalesced organizations of values, motives, experiences, aims, etc.). In most of us, however, one does not encounter coalesced, dissociated psychic trends even temporarily replacing the usual self-organization but, rather, indirect evidence in behavior of unconscious and disavowed ideas, aims, expectations, etc. In these more garden variety cases, there is no use served, other than confusion, to posit multiplicity of selves (save when there is no mistaking the metaphorical use of such concepts).

Consider the case of a woman, having recently given birth to a child, and now plagued with obsessive, unbidden thoughts concerning possible harm coming to her newborn infant. Here there is no question that the unbidden thoughts are consciously experienced. But they are experienced as unbidden, ego-alien, quite contrary to the person's consciously avowed plans, values, intention, feelings – i.e., contrary to the person's self-organization. Claparède (1951) has described how in certain cases of brain damage (e.g., Korsakow's Syndrome) memories lose their quality of 'me-ness'. Similarly for some cases of unbidden obsessive thoughts. In a certain sense, they are not mine. One of the ways in which they are not mine is that while I *have* these thoughts, I do not *think* them. Unbiddenly and passively, they occur to me. They are not mine in the sense that they are contrary to my conscious self-organization. Still another related sense in which these thoughts are not mine is that the wish attached to these thoughts is unconscious and disavowed.

Note that quite apart from the issue of unconscious aims, conscious experiences which I clearly have can, nevertheless, be experienced as not mine (in Sullivan's 1965 terms, 'not-me'). Rendering psychic trends within one's personality unconscious is only one means of disavowal. As Klein (1976) points out, one can also fail to understand the personal significance of certain experiences and, as I have tried to show, one can experience certain thoughts (and wishes) as not mine. In Frankfurt's (1976) terms, these thoughts, because they are rejected by the person, are made *external* to the person and according to Frankfurt, therefore, not attributable to him, "even though they may well persist or recur as an element of his experience" (p. 250). One should note here the similarity of Frankfurt's making external to Freud's suggestion that the experience of an originally inner event as external, as out there, is the original primitive defense (the 'ur-defense') and the model for all later defenses.

This example of unbidden thoughts warrants still a closer look. I have noted that the obsessive thoughts are too contrary to one's sense of personal identity to be experienced as 'mine'. For the new mother to acknowledge that she has desires to harm her infant would be too anxiety provoking and too overwhelmingly threatening to her sense of who she is. The solution is to disavow these thoughts (and desires) by rendering them external to her. What is important to note here is that such disavowals are self-preservative maneuvers designed to maintain the unity and integrity of the organization that already has been endorsed as self. To endorse thoughts and desires to harm her child would threaten to disrupt the integrity of what has been established as self. But, as the development of the symptom indicates, disavowal maneuvers do not necessarily work. The disavowed contents "persist or recur as an element of . . . experience." Furthermore, a price is paid in the use of disavowal. For one thing, disavowing and rendering certain contents as external to oneself increases the likehood that such contents will remain unintegrated into one's self-organization. As Charcot and Janet already stressed, the presence of dissociated, unintegrated contents weakens personality structure. It is the very opposite of the smooth functioning one observes when disparate aspects of the personality have been harmoniously integrated into the total structure.

Finally, dissociated, unintegrated material remains primitive and rigid, not subject to the corrections of experience, conscious deliberation, and problem-solving process. (Again, it is worth observing that the above 'prices' paid by disavowal were all noted by Charcot and Janet.)

THE USE OF 'I'

The above examples, in particular, their illustrations of disavowed engagements, should lead to a closer look at some of our habitual locutions.

When I say that "my behavior indicates that I process information, decipher meanings, have expectations, pursue aims, and follow rules of which I am not aware", I have used the same word 'I' twice. This may lead to the mistaken assumption that the word has the same reference and the same meaning in both uses. It seems to me that in saying "I have expectations of which I am unaware', I am using the word 'I' in two different senses. In the latter context, I am using the word 'I' to refer to that part of my personality, my conscious self, of which I am consciously aware and which constitutes my self-organization. In the former context, I am using the word 'I' to refer to my entire personality, including all the wishes, aims, expectations, values, etc., that are revealed by the entire range of my behavior. One can formulate this state of affairs in a highly unsatisfactory manner by saying that the expectations, aims, etc., of which I am not aware belong to another (unconscious) mind or another self. Or, the preferred alternative, one can conclude that the 'I' of "I am not aware" refers only to a limited aspect of any personality, the aspect available to conscious experience, and avowed self-organization.

Now what follows from the latter is that who I am is not limited to my conscious experience; my wants, aims and expectations are not fully revealed by what I experience wanting or expecting or what I say (to myself as well as to others) I want and expect. This is precisely what is implied, not only by Freudian theory (which elaborates this point with its partitioning schemes – its id, ego, and superego – and concepts of repression and the dynamic unconscious), but by a wide range of personality theories which distinguish between those aspects of the personality which are avowed by one's self-organization and the total personality. For example, in the personality theories of both Sullivan (1965) and Rogers (1951), the self-system is viewed as only one part of the total personality – a part limited to experiences, aims, etc., avowed and endorsed by the self. Experiences not in accord with the self-system are not repressed, as in Freudian theory, but not attended to and not fully articulated. In these theories too, who I am, the full range of what I experience, what I desire, and what my aims and goals are, are not fully revealed by my articulated, readily reportable conscious experiences.

ADAPTIVE FUNCTION OF SELF-ORGANIZATION

In the above discussion of the relationship between the self-organization and the total personality, the development and existence of the former has been taken for granted. One must remember, however, that self-organization and personal identity have evolved biologically as *adaptive* structures serving vital functions. It is likely that as complex a system as a person would require a superordinate structure whose main functions would include the coordination and fulfillment of a wide range of the person's interests and needs. Ideally, the wants, needs, aims, etc., of the self-organization would be fully congruent with those of the total personality. Or, to state it somewhat differently, ideally the pursuit of consciously experienced wants and aims would be fully congruent with one's organismic needs. The pursuit of wants and aims experienced as self-maintaining and self-enhancing would be a highly efficient way of meeting the full range of one's organismic needs. And *that* would be the main adaptive function of the self-organization. Now, it follows that a self-organization 'surrounded' by unintegrated, dissociated contents – that is, a self-organization which excludes a wide range of strivings and aims – does not adequately reflect the full range of the entire personality's interests and needs, and is therefore carrying out its adaptive function poorly. In short, in such a situation, *I* (defined as conscious self) am not adequately taking care of *my* (defined in terms of the entire personality) interests and needs.

That I may fail to endorse and avow which would be in my best intersts to endorse and avow once again suggests that the self-organization is not to be equated with the entire personality. Thus, like all or most human beings, one is likely to have, for example, sexual needs and the need for social-affective interaction with others. But one's personal identity and self-organization may be such that one does not endorse these needs or certain aspects of these needs, nor does one acknowledge either the focal or fleeting experiences which testify to these needs. Indeed, one can reject them and make them external to the self. As far as self-organization is concerned, one can say that one neither has these needs nor the associated desires. According to Frankfurt (1976), such desires and needs are "no longer to be attributed strictly to me . . . " (p. 250). But, in an important sense, from the point of view of my totally constituted personality, I continue to have these needs and desires – however rejected, dissociated, and externalized they may be. Frankfurt's analysis is cogent, but he ends the story too soon. For however rejected and dissociated, these desires continue to be mine in the important sense that they are part of my personality. It is important to understand that when one says that rejected and externalized desires are "no longer to be attributed strictly to me", the 'me' referred to is my endorsed self-organization but not my total personality.

What is touched on, by implication, in the above discussion is the whole issue of human nature. One can assume that certain desires, whether or not endorsed by the self, are universal and, therefore, part of every person's personality, even though not necessarily part of everyone's self-organization. It will be noted that this is essentially the assumption made by Freud with regard to sexual and aggressive urges. Whether or not one agrees with his specific formulations in this area, I believe he was essentially correct in believing that biologically laid-down imperatives are the best candidates as the source of such universal desires.[6] Also, I think it is important to recognize that, in part, the self-conception and self-organization one developmentally achieves is the product of an interplay between universal desires and one's particular history which leads one to endorse one set and one version of these desires and to reject another set and another version.

COHESION OF SELF

In most discussions, unity of self is understood in such a way that its logical contrast is multiple selves. The main question asked is whether one needs to posit multiple or mini-selves to account for psychic trends dissociated from one's conscious self-organization. The focus of attention has been directed, to use psychoanalytic terminology, to conflict between different psychic structures, that is, between self-organization and other trends in the personality. What has been taken for granted, until recently is, to begin with, the very existence of a self-organization. We have contrasted unity of self with multiple selves, but not with no cohesive self at all. The fact is, however, that the idea of a self or person entails some minimal degree of integration of disparate aspects of the personality. Consider, for example, a hypothetical case in which the informational input from the different sensory modalities were very different or even mutually contradictory. Without some integration of these incompatibilities there would be serious question regarding the integrity of one's personality or the unity of self. And note that here there is no question of multiple selves or of each mini-self subserving a different sensory modality. Rather, the problematic status of unity of self in this hypothetical case derives from the lack of integration among different functional systems within the personality. Here, divisions in the self or personality would refer to the unintegrated nature, the lack of coherent organization among these different functional systems.

Thus far, I don't believe that there would be any particular conceptual difficulty or controversy regarding the above description. One need merely accept the idea that in order to qualify as a person or self some minimal degree (admittedly unspecified) of coherence among our values, perceptions, wishes, goals, aims, etc., is necessary. It is not that failing this minimal degree of coherence we are beset with multiple selves but, rather, we can be said to have no self at all. At this level, the issue is not unity versus plurality but self versus no self.

The point to be noted here is that from a clinical and subjective point of view, the existence of wildly conflicting and dissociated wishes, aims, etc. is experienced, not as the possession of multiple selves, but as a threat to *any self or identity*. Colloquial expressions such as 'falling apart', feeling 'torn', 'splintering', 'breaking down' and other similar locutions capture this sense of threat to self-integrity and self-unity. It appears to be in the nature of having a subjective sense of self that it must be experienced as a numerical unity, as only one. Even a person with so-called multiple personalities would experience a single identity at any particular time or, if he were aware of his different personalities, would still say "*I* have different selves", thereby referring to a single superordinate self which suffers from this peculiar condition.

The psychological or subjective alternative to having a sense of self is not so much having a sense of two or more selves but a sense of no self or shaky or diffuse or even disintegrated self. What, from an external perspective, may be described as multiple selves, from a subjective perspective is experienced as a threat to any sense of self. (The person saying "I have two selves" is taking an external perspective in relation to his multiple selves.)

Consider the symptom of depersonalization in which one experiences oneself as out there observing oneself. From an external perspective, one can describe this phenomenon as both an observing self and a self being observed as object. But the person experiencing severe depersonalization experiences an intensely anxiety-provoking experience of a disintegrating self.

For most of us, who have achieved at least a minimal degree of coherence necessary to a sense of self, self-unity is maintained partly through the dissociation and disavowal of aims, wishes, desires, etc. markedly incongruent with one's sense of who one is. That which is 'ego-alien' is expressed indirectly and in disguised form (as in symptoms and dreams). And these disguised unconscious and disavowed aspects of the personality have been the main focus of traditional psychoanalytic theory.

Much recent psychoanalytic literature, however, has concerned itself less with neurotic intrapsychic conflict – the original arena for the anatomy of the personality and more with developmental impairments in self-cohesiveness found prominently in so-called narcissistic personality disorders and borderline conditions. Such impairments are described in terms of such dimensions as failure to differentiate fully between self and other and the use of 'splitting' as the characteristic means of dealing with incompatible affects and valuations. In general, the description in these cases is of an inadequately developed self rather than, as in neurosis, conflict between a relatively coherent self-organization from which incompatible aims, strivings, etcetera, are dissociated and disavowed. As Kohut (1977) puts it, while in neurosis, what is primary is conflict among relatively intact structures, in narcissistic personality disorder and borderline conditions, the primary problem is failure to achieve an intact self-organization. This is seen clinically in, for example, the susceptibility of the latter to experience what Kohut refers to as "disintegration anxiety" – an experience of disintegration and dissolution of self which derives from a lack of intact sense of self.

The point to be noted here is that while the traditional concern has been with those threats to unity of self and personality which are linked to 'double-mindedness' (Kierkegaard, 1956), "bad faith," (Sartre, 1956) and 'intrapsychic conflict' (Freud, 1917) a basic threat to the very existence of self derives from developmental 'weakness' in the self structure.[7]

SELF-COHESIVENESS, IDENTITY OVER TIME, AND IDENTITY AT ONE TIME

As noted at the beginning of this paper, until recently for the most part, philosophers have tended to concern themselves more with personal identity *over time* than with the issue of personal identity 'at a time' (Perry, 1975). I want to try to show briefly, through a further discussion of self-cohesiveness, the relationship between these two aspects of personal identity.

I mentioned above that 'splitting' is employed by borderline patients as a characteristic means of dealing with incompatibilities. This may be worth pursuing for it will help demonstrate that the issues of personal identity over time and identity at one time converge at certain points. Prominent in clinical descriptions of splitting are radical alternations in behavior such that the person is all-loving and all-idealizing toward another one one occasion and all-hating and all-denigrating on another occasion. Such a radical shift over different occasions obviously reflects instability of self over time, but it also reflects lack of unity *at one time* in the following senses: One assumes that every adult person in a close interpersonal relationship will have complex affective reactions to the other, reactions which will broadly include elements of love and hate, frustration and gratification, disappointment and satisfaction, idealization and denigration. This is, of course, essentially what is referred to when one speaks of an inevitable degree of ambivalence in close relationships. In other words, all these complex and conflicting reactions and valuations will be part of one's *ongoing and current* set of feelings toward and images of the other and will be integrated into some total durable affective-cognitive stance toward the other and durable image of the other. The borderline person, however, responds to these integrative demands by dissociating positive and negative affects and valuations from each other so that they alternate *temporally.*[8] In important respects, he functions like a multiple personality. The differences lie in the fact that the borderline person remembers state 2 while in state 1 (or vice versa) and that the radical alternatives seem to be limited to affects and valuations in close relationships and do not extend to total personality strivings, aims, styles of behavior, etc.

Both the borderline person and the multiple personality deal with *current* conflicting feelings and valuations by *alternating them temporally*. What is clearly implied here is that the feelings and valuations that become manifest in state 2 are also present, in a dissociated form, when the person is in state 1. It is in this sense that the issues of stability over time and unity at one time converge.[9] Put very simply, radical instabilities over time are to be viewed as an expression of failure of unity at one time. It is not only philosophers who wonder whether one has the right to see someone who has shown radical instability over time as the same person. The person himself – through his anxiety, and feelings of 'disconnectedness' symptoms – may be revealing to us that he is experiencing the same doubts.

SPLIT-BRAIN PHENOMENA

Until this point, I have said little of recent findings with split-brain patients. I had wished to focus on the challenges to unity of self represented by psychoanalytic formulations and the clinical phenomena which prompted these formulations. These recent findings, some believe, constitute perhaps the most serious challenge to a simple, traditional conception of unity of the self. Consider the following reports with split-brain patients.

(1) Two different tasks which, if directed to one hemisphere, would create interference, can be carried out by the two hemispheres simultaneously with little interference (e.g., Gazzaniga and Sperry, 1966).

(2) The right hemisphere can process information and initiate behavior independent of and in ways unknown to the left hemisphere (e.g., Sugishita, 1978).

(3) The left hemisphere reacts with appropriate emotion to information presented to the right hemisphere even though the left hemisphere does not known the context or perceptual nature of the information (e.g., Gazzaniga *et al.*, 1977).

(4) Despite its ignorance, the left hemisphere provides a plausible explanation for the emotional reaction and behaviors emanating from the right hemisphere (Gazzaniga *et al.*, 1977).

These findings appear to challenge any simple conception of unity of the self insofar as (1) sub-systems (i.e., each hemisphere) appear to be capable of the kinds of complex and intelligible activities which we have tended to assign only to a person and (2) the sub-systems can carry out these activities somewhat autonomously of each other. (It will be recalled that to Charcot and Janet it was the *autonomous* operation and development of certain ideas and affects which was held to be the critical feature of dissociated states such as multiple personalities, automatic writing, hysterical symptoms, and hypnotic phenomena.) Although one should be cautious in trying to extend findings from commissurotomized patients to intact individuals, it is tempting to speculate that in patients showing extreme dissociated states there may be functional failures of integration between the two hemispheres. Also, it is likely that for all of us, unity of personality is, in part, a function of the integration and harmonious interaction between the two hemispheres.

What can be said of the two hempispheres would seem to be applicable, however, to the operation of any identified functional systems. That is, unity of self and of personality are best conceptualized in terms of integration, not only between the two hemispheres, but among a wide range of the different functional systems comprising a person. Thus, the failure to integrate, let us say, an important memory sub-system (as in cases of amnesia or Korsakow's syndrome – see Claparède, 1951) is likely to be at least as disturbing to unity of self and personality as failure to integrate the functioning of the two hemispheres. Indeed, the clinical dissociated states referred to above are perhaps more readily understood in terms of unintegrated memory, cognitive, and affective sub-systems or structures rather than in terms of failure of inter-hemispheric integration. In any case, it seems to be that 'unity of self' and 'unity of personality,' are, on a psychological level, the terms we use to refer to the range and degree of functional integration of various sub-systems comprising the individual.

What has made the split-brain phenomena such fertile ground for talk about multiple persons and minds is that the failure of integration (between the two hemispheres) is patent and dramatic. But two considerations need to be kept in mind. One is that even split-brain patients, for the most part, show coordinated and integrated behavior. Most often, it takes special laboratory techniques to reveal their special difficulties. The second point is that many of us show failures of integration in the arena of emotional conflict which, while perhaps more hidden and less dramatic than split-brain patients, may have as profound an effect on one's personality. But there too, we show large areas of integrated behavior. Were there little or no integration, we probably would not be entitled to think of ourselves as persons. Similarly, were the split-brain patients' behavior completely dominated by contradictory and unintegrated left and right hemisphere activities, one would have to question seriously whether one is talking about a single person or perhaps a person at all. But, as noted earlier, this question can and should be directed to the relationship among any of the functional systems comprising the personality. As Robinson (1976) points out, behavior more subtle than, but similar in important respects, to split-brain phenomena can be observed in all of us. The basic point perhaps to be made is that to the traditional forensic criteria of personhood of a certain minimal level of cognitive competence and self-reflective capacity one should add the criteria of a minimal level of coherence and integration among component functions and processes. In this sense, person is a hierarchical concept involving a superordinate organization and integration of component functions and processes.

Before leaving the split-brain findings, one final point regarding the role of linguistic rationalizing processes in maintaining a sense of unity. As noted above, the left hemisphere provides a plausible explanation for the emotional reactions and behaviors emanating from the right hemisphere even when the former does not know the perceptual basis for these reactions and behaviors. LeDoux (1978) and Gazzaniga *et al.*, (1977) suggest that these "verbal attribution processes" are the left hemisphere's way of maintaining a sense of self unity in the face of a situation in which it has emotional reactions without knowing what it is reacting to and shows behaviors without knowing their basis. This raises a number of interesting questions which can only be noted and not pursued here: for example, to what degree do the motive and reason explanations of intact individuals consist of rationalizing verbal attribution processes which help maintain the comforting and self-unifying illusion that we always (or most frequently) know precisely why we do what we do and feel the way we do? (See Eagle, 1977; Nisbett and Wilson, 1977.) The use of verbal attribution processes is comforting and self-unifying, not only insofar as it provides *some* reason for doing and feeling X (a sense that *I* am doing and feeling X), but also because of the *specific nature and content* of the reasons offered. That is, the reasons and motives we give are likely to be consistent with the values, beliefs, images, prohibitions, etc., held by the conscious self and hence, serve to maintain the self structure. In other words, we maintain self-unity, not only by disavowing impulses and wishes inconsistent with the conscious self, but also by disavowing reasons and motives inconsistent with the conscious self and offering instead reasons and motives consistent with the self-system.

CONCLUSION

Even if one accepts the assumption of numerical unity, one is still left with the question of degree of *integration* of a particular self or personality. And as I have tried to show, it is this dimension, rather than numerical unity, which is the critical one in dealing adequately with the unity of self and unity of personality issue. We may agree, for a wide range of people, that each one is best thought of as a single person and a single self. But a single person can be integrated and unified or nonintegrated and nonunified and a single self may be cohesive or fragmented. The main position I want to urge is that while most of us may be assured of numerical unity (that is, one self per body), as far as the issue of integration is concerned, unity of self and of personality is a developmentical integrative achievement in which we are not all equally successful. All the clinical phenomena with which psychoanalytic theory is concerned and with which prepsychoanalytic dynamic psychiatry was concerned, to quote Ellenberger (1970) " . . . illustrate the fact that unity of personality is not given to the individual as a matter of course, but must be realized and achieved through the individual's persistent, and perhaps life-long efforts" (p. 141).

York University

NOTES

[1] As Ellenberger (1970) points out, Janet had also spoken of personal motivations for dissociated states. But Janet did not fully develop this idea nor did he give personal motivations the central role they came to assume in psychoanalytic theory.

[2] As I will try to show later in the paper, more recent psychoanalytic literature (e.g., Kohut, 1971; 1977) has been more concerned with unity of self – at least, cohesiveness of self – than with conflict or lack of conflict among different components of the personality.

[3] In the case of split-brain patients, this mind and this conscious self is the one associated with the left hemisphere.

[4] Entirely conscious conflicts, however, are more likely to be resolved through the operation of rational devices available to consciousness. What renders unconscious conflicts more recalcitrant is, among other things, the fact that unconscious wishes and aims remain unintegrated, not subject to the problem-solving and integrative operations of consciousness.

[5] Historically, the terms 'id and 'ego' have carried two somewhat disparate sets of meanings in psychoanalytic theory. In one set, which I am stressing here and which is often lost sight of in the English translation, 'id' refers to an impersonal 'it', while 'ego', best translated as 'I', denotes that which is owned and, so to speak, autobiographically experienced. In the other set of meanings, which has dominated psychoanalytic thinking, 'ego' is defined primarily in terms of certain (reality-testing, delaying, and defensive) functions, while 'id' has been equated with instinctual drive (sex and aggression in Freudian theory). Now, as long as one assumes that by their very nature instinctual derivatives will necessarily constitute the predominant portion of what is repressed and

dissociated, the conflation of these two sets of meanings will not be problematic. If that which is experienced as an impersonal and 'ego-alien' "it" necessarily consists primarily of instinctual drive material, – that is to say, if they are equivalent – then it matters little whether 'id' is defined primarily as impersonal "it" or in terms of instinctual drive. And, of course, Freud assumed a general equivalence between instinctual drive and that which is repressed and dissociated. If one believes, however, that the repressed and dissociated need not be limited to sex and aggression, the conflation of the two sets of meaning does become problematic. As Moore (1980) notes:

The universal, functional definition of the id will simply get in the way of the personally variable parts of the total personality that each person disavows or represses. It simply will not be the case universally, for example, that a person's sexual wishes will either be repressed or disavowed in some other way. Sometimes that will be the case and sometimes it will not. Thus, when the id is used as a synonym for that part of the total personality not affirmed as part of oneself, sometimes such sexual wishes will be included in the id, and sometimes they will be excluded. The problem lies in the fact that this highly individualized mode of aggregating the mental states belonging to the id will not match the universal attribution of those sex drives of any person to the id demanded by Freud's functional organization of the person. . . . This lack of congruence between the id defined as the excluded total personality and the id defined by its functional organizing principle will result in two senses of "id", just as it will in two senses of 'ego'. The result of such two quite different things being named by the same name is confusion, not theoretical advancement. (p. 44–5)

Although it obviously cannot be fully discussed here, a basic issue that Freud raises by his conception of id as instinctual drive is the psychological status of biological imperatives. That is, if one assumes, with Freud, as well as with the ethologists and sociobiologists, that we are born with certain genetically programmed instinctual imperatives, questions arise regarding their psychological and subjective status. Are they indeed, psychologically represented? If so, how? As urges, desires, and, to use McDougall's (1922) term, sentiments? For example, if Dawkins (1976) is correct and a good deal of all animal and human behavior is functionally related to the aims of preserving and transmitting one's genes, how is this aim represented psychologically? By such diverse institutions, behaviors, and feelings – most of which, it should be noted are culturally based and sanctioned – as family structure, parent-child attachment, sexual desire, and feelings for one's kin?

Freud it seems to me, was never clear whether the id belonged to the repressed and impersonal because of the prohibitions of civilization or because of its very nature as a primal inchoate "ever-seething cauldron" (Rapaport, 1954). One can find evidence for both points of view. (Anna Freud, (1966) makes clear that she opts for the latter interpretation when she speaks of the "ego's primary antagonism to instinct" (p. 157).) That is, at times Freud seems to be saying that the id is never fully represented psychologically or subjectively because it is essentially a biological concept (or, at best, on the borderland between the body and the mind); at other times, he seems to be saying that it is relatively unrepresented psychologically because, given a particular life history and certain social prohibitions, it is subject to repression. It is this duality of usage to which Moore refers in the above quoted passage.

The use of id in terms of repression and dissociation of what is radically contrary to how one defines oneself is relatively unproblematic (or, at least, no more problematic than the general idea that one may disavow and be unaware of certain of one's strivings, aims, and wishes). In this use, the content of what is repressed and dissociated would be determined by the particular nature of one's personal history. But the concomitant definition of id as quasi-biological drive and the belief that *as biological drive* instinctual urges are necessarily dissociated – that is, are to be contrasted with and conceptually separated from the realm of that which is experienced as personal (as 'Ich') seems to me to involve a confusion of levels of discourse. That is, quite obviously biological drives and the physiological processes underlying them are not directly experienced. Biological imperatives are expressed and revealed in certain feelings, aims, and desires. Thus, lust rather than hormonal secretion and anger rather than hypothalamic stimulation are experienced. But this is also true of the functioning of, let us say, one's circulatory system or central nervous system. One experiences energy versus fatigue, not the operation of one's circulation; and one experiences particular percepts, not neurons firing. One does not refer to the 'imperatives' of one's circulatory system or one's central nervous system as a separate domain divorced from one's personal self simply because the former refers to a different level of analysis and discourse than the latter. Similarly, why should one refer to the biological drives of sex and aggression, *qua* biological drives, as a sub-structure of the personality, to be contrasted with the realm of personal experience? Only if one wants to also define ego in terms of biological functioning can one meaningfully contrast id with ego. Or alternatively, only if one wants to talk about both id and ego in terms of a single level of discourse – as dissociated versus avowed aims and strivings, for example – can one meaningfully contrast them.

It should be noted that sex and aggression have no special status with regard to susceptibility to disavowal and dissociation. One can deny fatigue, disavow a political belief, or a percept, or reveal in one's fantasies and behavior dissociated wishes to merge with another. The point then, is that to say that "where id was, there shall ego be" cannot sensibly mean that where biological drives were personal experience shall be. (Just as one could not sensibly say that where circulation was, there shall direct experience of fatigue or energy be; or where firings of neurons were, there shall the direct experience of the percept be). It can only mean that where the dissociated impersonal 'it' was, there shall experiences and aims which are personally owned and avowed be. And if this is so, one must also acknowledge that the particular content of what is dissociated impersonal 'it' – that is, the particular content of what needs to be owned, avowed, and integrated – must be empirically determined by the histories and experiences of particular individuals.

(It is worth mentioning, in passing, that Freud's assumption of the equivalence between the instinctual and that which is repressed and dissociated – an assumption which lies at the heart of his instinct theory – has represented, from prepsychoanalytic thinking to current psychoanalytic developments the essential basis for divergence from Freudian theory (e.g., see fairbairn, 1952; Kohut, 1971, 1977).

[6] That certain biologically grounded wishes and desires are universal (e.g., sexual desires) is one basis for Freud's assumption that even when these desires are not consciously experienced they, nevertheless, exist in some unconscious, repressed form. However, as argued in note 5, even if Freud is correct, this would not mean that the class of that which is repressed is equivalent to the class of universal desires. For one, as Moore (1980) points out, not everyone represses these universal desires; and two, as I have

argued, certain wishes and aims which do not belong to the class of universal desires can also be repressed.

[7] Note the similarity between this current view of certain classes of psychopathology and the belief of nineteenth-century psychiatry that certain patients were constitutionally weaker in their capacity to integrate disparate trends in the personality and hence, were more likely to exhibit dissociative phenomena. Substitute developmental difficulties for constitutional predisposition, and there is not too great a difference between the two formulations.

[8] In a certain sénse, what is involved here is not only unity of the self, but unity of the *other*. In the course of development, one normally achieves not only a stable self-organization, but a stable image of the other – what has been referred to as object constancy. What one sees in 'splitting' is an impairment in stability of both self and other.

[9] There is another, more subtle, sense in which the two issues converge. In Freudian theory, it is the residue of infantile wishes, aims, and striving from the *past* that are most difficult to integrate into one's self-organization. Indeed, neurosis can be seen as the partial failure of such integrative attempts. In a certain sense, one can say that neurosis consists in a dissociation of the past which causes radical discontinuities over time. And therapeutic growth consists partially in a rediscovery and avowal of the past and integrating its current residues into adult self-organization. The convergence of stability over time and unity at one time consists in the facts that a viable self-organization requires the integration of past into present and that experiences of instability over time, of discontinuity with the past are, therefore, often experienced as threats to *current* identity and unity of self.

REFERENCES

Armstrong, D. M.: 1968, *A Materialist Theory of the Mind*, Routledge and Kegan Paul, 1968. Academic Press, New York.

Breuer, J. and Freud, S.: (1893–1895) *Studies on Hysteria*, SE, Vol. 2. Hogarth Press, London.

Claparède, E. 'Recognition and 'me-ness'.' in D. Rapaport (ed.), 1957, *Organization and Pathology of Thought*, Columbia University Press, New York, pp. 58–75.

Dawkins, R.: 1976, *The Selfish Gene*, Oxford University Press, New York.

de Sousa, R.: 1976, 'Rational homunculi', in A. O. Rorty (ed.), *The Identities of Persons*, Univ. of Calif. Press, Berkeley and Los Angeles.

Eagle, M. 'A critical examination of motivational explanation in psychoanalysis', Paper given at the Center for the Philosophy of Science, University of Pittsburgh, November 1977. (Also to appear in: A Grünbaum and L. Laudan (eds.), *University of Pittsburgh Series in Philosophy of Science*. Berkeley and Los Angeles, University of California Press; and *Psychoanalysis and Contemporary Thought*, (in press).

Ellenberger, H. E. 1970, *The Discovery of the Unconscious*, Basic Books, Inc., New York.

Fairbairn, W. R. D.: 1952, *Psychoanalytic Studies of the Personality*. Tavistock Publications, London.

Fehrer, E. and Raab, D.: 1962, 'Reaction time to stimuli masked by metacontrast' *Journal of Experimental Psychology* **63**, 143–147.

Field, G. C., Aveling, R., and Laird, John: 1922, 'Is the conception of the unconscious of value in psychology?' *Mind* **31**, 413–442.

Fingarette, H.: 1969, *Self-Deception*, Humanities Press, New York. (Also Routledge and Kegan Paul, London.)

Frankfurt, H.: 1976, 'Identification and externality', in A. O. Rorty (ed.), *The Identities of Persons*, University of California Press, Berkeley and Los Angeles.

Freud, A.: 1966, *The Ego and Mechanisms of Defense*, (Revised Edition). International Universities Press, New York.

Freud, S.: (1900), *Interpretation of Dreams*, Standard Edition, Vols. 4 and 5, Hogarth Press, London.

Locke, J.: 1894, (1689), *Essay Concerning Human Understanding*, Clarendon Press, Oxford.

McDougall, Wm.: 1922, *An Introduction to Social Psychology*, Methuen, London.

Moore, M. S.: 1979, 'Responsibility for unconsciously motivated action', *International Journal of Law and Psychiatry* **2**, 323–347.

Moore, M. S.: 1980, 'The unity of the self', Paper given at Third International Conference on History and Philosophy of Science, Montreal, Canada.

Nagel, T.: 1973, 'Brain bisection and the unity of consciousness', in J. Perry (ed.), *Personal Identity*, University of California Press, Berkeley and Los Angeles. Also in *Synthese* **22**, 1971.

Nisbett, R. E. and Wilson, T. D.: 1977, 'Telling more than we can know: Verbal reports on mental processes'. *Psychological Review* **84**, 231–259.

Perry, J. (ed.) 1975, *Personal Identity*, University of California Press, Berkeley and Los Angeles.

Puccetti, R. 1973, 'Brain bisection and personal identity'. *British Journal for the Philosophy of Science* **24**, 339–355.

Raab, D. 1963, 'Backward masking', *Psychological Bulletin* **60**, 118.

Rapaport, D. 1954, 'The autonomy of the ego', in R. P. Knight and C. R. Friedman (eds.), *Psychoanalytic Psychiatry and Psychology*, International Universities Press, New York.

Robinson, D. N.: 1971, 'What sort of persons are hemispheres? Another look at split-brain man', *British Journal for the Philosophy of Science* **27**, 73–79.

Robinson, D. N. 1971, 'Backward masking, disinhibition and hypothesized neural networks', *Perception and Psychophysics* **10**, 33–35.

Rock, I.: 1979, 'Perception from the standpoint of psychology', in Research Publication, *Perception and Its Disorders*, Association for Research in Nervous and Mental Disease, 68, 1–11.

Rogers, C. R.: 1951, *Client-Centered Therapy*, Houghton Mifflin Co., Boston, pp. 481–533.

Sartre, J. P.: 1956, *Being and Nothingness*, Hazel E. Barnes, (Transl.) (Chapter 2 Mauvaise Foi and The Unconscious). Philosophical Library, New York.

Schafer, R.: 1976, *A New Language for Psychoanalysis*, Yale University Press, New Haven and London.

Siegler, F. A.: 1967, 'Unconscious intentions', *Inquiry* **10**, 251–267.

Sperling, G.: 1963, 'A model for visual memory tasks', *Human Factors* **5**, 19–31.

Sugishita, Morihiro.: 1978, 'Mental association in the minor hemisphere of a commissurotomy patient', *Neuropsychologia* 16, 229–232.
Sullivan, H. S.: 1965, *Collected Works*, Norton, New York.
Thalberg, I.: 1977, 'Freud's anatomies of the self', in R. Wollheim (ed.), *Philosophers on Freud*, Jason Aaronson, Inc., New York.
Williams, B. A. O.: 1973, *Problems of the Self*, Cambridge University Press, Cambridge.

MICHAEL S. MOORE*

THE UNITY OF THE SELF

Michael Ruse (ed.), Nature Animated, 163–202.

1. INTRODUCTION

It has long been thought that the findings of psychoanalysis challenge the common sense idea of the unity of the self. As is well known, Freud produced two subdivisions of the self as the best conceptualizations of the data of psychoanalysis, first, the topographical division in terms of the "systems" conscious, preconscious, and unconscious, and later, the structural subdivisions in terms of ego, id, and superego. In producing some such subdivisions, Freud was not without impressive forerunners, including Plato's tripartite division of the soul, Hume's separation of reason from the passions, or Nietzsche's Dionysian versus Apollinian characters. Unique to Freud's theory, however, is the seriousness with which it attributes actions, mental states, awareness, purpose, character structure, and sometimes even responsibility, to these subdivisions of the mind. These "characters" are given attributes usually reserved for whole persons. Hence, the fracturing done by psychoanalytic theory to the unity of the self seems to be much more serious than that of its predecessors.

The present paper will analyze the claimed disunity of self of psychoanalytic theory in four steps: (1) a survey of our moral, legal, and metaphysical needs for some doctrine of a unity of self; (2) an ordering of the data of psychoanalysis and of the data related to multiple personalities, split brains, and some of the more recherché thought experiments of the contemporary philosophy of self-identity, by their seeming ability to challenge the assumptions of unity of the self; (3) an analysis of the various senses one might assign to the idea of "the unity of the self," and an analysis of the degree to which the data of disunity point toward disunity in any of such senses; and (4) to the extent that the data reveal disunities of the self, an analysis of whether the metapsychological viewpoints of psychoanalytic theory are the best way to conceptualize such disunities. My conclusions, simply put, are that our needs for a concept of a unified self are very basic and it would be hard to even imagine giving it up; that nothing in psychoanalytic theory or related theories can show a disunity of self in any very serious sense of disunity; and that to the extent that there are disunities of the self to be talked about, that neither the structural nor topographical metapsychologies is the way to talk about them.

2. OUR NEED FOR SOME CONCEPT OF THE UNITY OF THE SELF

Broadly speaking, one should distinguish two sorts of needs we have for the concept of the unified self, our metaphysical needs for such a concept, and our moral and legal needs for it.[1] I shall discuss each in turn.

2.1. *Metaphysical Needs for the Unity of the Self*

2.1.1. *The Need for Parallel Individuation of Mental States and Physical States*

'Metaphysical' is not here used in a perjorative sense. All that is meant by the word in this context is to identify those needs we have that reflect our most general and abstract aspects of ourselves. There appear to be at least four aspects to our nature that seem to demand or make important that we have some concept of the unity of the self. The first stems from the way in which we individuate mental states on the one hand and physical states, on the other hand. We individuate mental states such as intentions, hopes, desires, fears, etc., not only by their contents and the time at which they are held, but also by the person who holds them. Two people who each desire that the very same state of affairs obtain at the very same time still have (numerically) distinct desires. By contrast, we individuate physical states – using the phrase very broadly to encompass dispositional states and functional states as well as straightforwardly physical states – in part by the physical body whose states they are.

If we are "disunified" in what I take to be the basic sense, viz, that there is more than one person per human body, then we may rule out any non-dualist solutions to the mind/body problem. Suppose, for example, there were three persons per body. A mental state that we normally would regard as one mental stake-token, we now must regard as three.[2] Yet the corresponding physical state with which we might hope to identify the mental state in question is only one state. This would seem to rule out any identification being made of mental states with physical states, and rule out, accordingly, the three most tempting metaphysical views on the relation of bodies and minds, namely, materialism, behaviorism, and functionalism.[3] Indeed, one would seem to be committed to some form of dualism, with all of the problems so well detailed in the philosophy of mind since Gilbert Ryle (1949), The way not to be saddled with dualist metaphysics is to adhere to the idea of "the unity of the self" in one of its most important senses, namely, that there is only one person to be found 'in' any given human body.

2.1.2. *The Need for Parallel Individuation of Bodily Movements and Basic Human Actions*

Analogous problems arise for our basic ideas about human actions if we begin to think of each human being being composed of multiple selves. How one individuates human actions has been a much debated matter in the contemporary philosophy of action. There are proponents of 'fine-grained' modes of individuating, proponents of 'coarse-grained' modes, and compromise positions.[4] Taken for granted by both sides of the contemporary debate is that one can individuate *basic* acts,[5] no matter how one comes out on identifying such basic acts with more complex acts. The way in which we individuate basic acts is in part by the person whose acts they are. If two different people perform the very same kind of basic act, those are two distinct acts. By contrast, the bodily movements that in some suitably loose sense "constitute" the basic acts,[6] are individuated by the body whose movements they are.

Again, contradiction seems imminent if we maintain these modes of individuating acts and movements *and* maintain that there is more than one person per human body. Suppose, for example, some action is a 'compromise formation' between the activities of the id and of the ego. There should be two basic acts by these 'people', but they have only one body through which they can act. By the bodily movement criterion, this should be one basic act, but by the personal criterion there should be two. The only way for a multiple selves theorist to avoid the contradiction would be to deny that there is any special relation between some particular raising of an arm and the basic act of raising that arm on that occasion. However one comes out on Wittgenstein's (1953) famous question of what is left over if we subtract the fact that my arm goes up from the fact that I raised it, it surely is not going to be that the movement and the act bear no intimate relation to one another.[7]

2.1.3. *The Epistemic Need for Unity*

The third of our metaphysical needs for a unified self stems from the inheritance of Descartes and Kant, namely, the need for unified *epistemic* self. In order to have knowledge of anything, the subject who perceives seemingly must be the same as the subject which draws inferences from such perceptions, who imagines what else could be the case, who remembers what was the case, and who draws conclusions from all of this and gains knowledge. If there are different persons who do each of these things, one has difficulty in grounding human knowledge, as Descartes recognized in his famous starting point of the 'I' in refuting skepticism. The "condition for the possibility" of knowledge may be that there be a unified self as the knower or the subject of knowledge.

2.1.4. *The Experience of Unity*

The fourth of our metaphysical needs stems from the experience we have of ourselves as being the same self both at a time and over time. We usually experience simultaneous perceptions, imaginings, emotions, intentions, inferences, and the like, as being *ours* (and thus of one person); moreover, we remember past experiences as being ours as well. Whether this unity of consciousness can be made into a necessary or a sufficient criterion for identifying persons has been a much debated matter since Locke.[8] To the extent that there is such an experience of unity – and some of the data of psychoanalytic theory rather directly challenges this – it surely at least inclines us to a view that there is at most one self 'in' our bodies at any given time.

2.2. *Moral and Legal Needs for the Unity of the Self*

2.2.1. *Persons as the Holders of Rights*

The most basic facts about our nature – our physical embodiment, our possession of consciousness, our being the subject of knowledge – all suggest some doctrine(s) of the unity of the self. In addition to these metaphysical needs for such a doctrine, our moral and legal theories seemingly presuppose the same doctrine. Consider first persons as the holders of moral or legal rights. Each person, for example, has a right to receive justice in the distribution of social goods. Any plausible theory of distributive justice will involve the idea of the equality of persons and their prima facie entitlement to an equal distribution of social goods. Such ideas make sense only if they presuppose some way of individuating persons, and the principle of individuation our sense of justice rather clearly employs is one person per body. Otherwise, for example, schizophrenics, multiple personality persons, and classical Freudians would all be able to use the car pool lanes reserved for more than one person on the freeways, even though they were, so to speak, 'by themselves'.

Our assignment of property rights also presupposes some unity of the self. Consider the ownership we each have of our own bodies, a property relation insofar as we can sell our hair, blood, donate our organs, exclude other persons from contact or intrusion. Suppose one is a Lockean about property rights, so that in the first instance such rights are acquired either by possession or by the mixing of one's labor with the property. If there is more than one self per body, this leads to some very strange results. Suppose, for example, that we were to think of a multiple personalitied person as being different persons at different times. Presumably the first person to 'possess' the body gains title to it on the occupation version of Locke's theoyr. Alternatively, on the labor theory, presumably the first self that improves the body in some way (a regimen of exercise, perhaps, or the acquisition of a sexual partner) gains the property rights. (Gluttonous selves would on either version be excluded from owning the body because of the Lockean proviso against waste.)[9] In any case, however title is acquired, one would want to know what would happen when the owner left. Should this be treated as an abandonment? Even if it is, perhaps there was fraud on the part of the second possession (i.e., self-deception) which would vitiate any purported abandonment. Alternatively, suppose the ownership relation continues after the first self leaves. Eventually his title would be lost because of the running of the statute of limitations for the recovery of personal property, because typically title is lost to another who takes adverse possession. Of course, one should not be too hasty here: for sometimes the statute of limitations may be tolled if the owner is out of the jurisdiction, which presumably the first possessing self was when another self took charge. Additionally, such a self might again raise a claim of fraud, a good defense to adverse possession, if there is (self-) deception with regard to the intent to remain in possession.

This is admittedly a rather silly story, but like most *reductio ad absurdum* arguments the point is to thrust the absurdity of the conclusion back on to the premises that generated it. Since neither the Lockean theory of property nor the idea of property rights in a body is *that* silly, that leaves the idea of more than one person per body as the suspect premise.

2.2.2. *Persons as the Subjects of Responsibility*

Our moral and legal theories not only assign rights to persons, they also ascribe responsibility. These theories too would be considerably different than they are if we could not assume the unity of the self. One can see this by adverting to the conditions under which responsibility is ascribed, in both morals and law. We hold someone prima facie culpable if he performs an action intentionally causing harm; we hold him actually culpable if, in addition, he lacked any of the justifications or excuses for actions that mitigate or eliminate responsibility.

Multiple selves would make a hash of the conditions under which we ascribe moral fault, for it would lead to contradiction at every turn. Multiple selves, some of whom are knowledgable and some of whom are ignorant, would result in actions being both intentional (for the knowing self) and unintentional (for the ignorant self) in every case; indeed, insofar as knowledge is the hallmark of basic actions at all, every basic act by some subagent would also fail to be a basic act by some other subagent of a multiple-personed body. Such contradictory implications for responsibility would also exist if one had to apply established excuses to responsibility to multiple selves. Consider one excuse, duress, the acting under threat by another. Given the dynamics of multiple selves Freudian theory posits in, say, obsessional neurosis, the result would be that one self would be acting under the threat of another self; in such a case, responsibility would be undeterminable, because if there is one self who is threatened and thus excused, there is another self who is the threatener and who would not be excused. So in excuses as in the prima facie case, there will be no answer to the overall question, was he (the whole human being) responsible, for in all cases he both was and wasn't.[10]

Because the ascription of moral fault is a necessary condition to the imposition of liability in criminal law, this quandry about moral responsibility would also be a quandry about legal liability. Unless one could devise punishments that punished only the guilty self in some body, multiple selves within the same body would face the legal system with the choice between a radically extended system of vicarious responsibility, or not punishing anyone ("better to let ten guilty selves go free than to punish one innocent self").

2.2.3. *Persons as the Subject of Self-Interest*

As a last example of the presupposition of the unified self of our moral and legal theories, consider the calculation of self-interest recommended by egoism. While egoism is not much of a *moral* theory, we all have a special concern for our own self-interest, and so will share any presuppositions of unified selves of this theory. If there are multiple selves sharing the same body, our idea of self-interest would be considerably different than it is. We conceive of self-interest in terms of all of our needs at a time. We arrive at our overall self-interest by taking into account all of the needs, desires and emotions that seem relevant to any particular decision. Yet if my ego, say, were a different self than my id, why should the one care about the other?[11] Presumably each – themselves being egoists – have that special concern only for their own self-interests.

Regarding some mental states as belonging to another would not reflect our experience in taking such interests into account as *ours* in framing our overall self-interest. We weigh such interests and form what Frankfurt (1971) calls "second-order desires" with regard to them, that is, we may discount some desires because we think them less worthy than others. This kind of merging of component wants into an overall want would be impossible if we have separate selves within the body. Rather, 'one' (whoever he is) would have to construct either a social welfare function for "inter-personal" utility comparisons, or adopt a scheme of distributive justice between such selves, in order to calculate what we call (overall) self-interest. Not only is the first no more possible for selves within a body than it is for separately embodied selves, and not only is the second inconsistent with egoism; but neither mode of taking into account the differing interests one may have on any particular occasion squares with our experience of forming such components into our own self-interest.

In each of these ways, our moral and legal theories demand that we regard ourselves as unitary. Not surprisingly, this presupposition matches that required by our most basic, metaphysical conceptions of ourselves. The single self presupposed by our being embodied, conscious, and knowledgeable is the same single self that can hold rights, be responsible, and maximize his own or others' self-interest. Psychoanalytic theory is as interesting as it is partly because it seems to challenge this very basic presupposition of our metaphysics and morals.

3. THE PSYCHOANALYTIC CHALLENGE TO THE UNITY OF THE SELF

A useful analytic device which separates the various factors suggesting disunity of the self, is to begin with a very simple example and add other factors one by one. This is done below, the ordering principle being my own intuitions that as the list progresses we are increasingly tempted to speak of more than one person per body. These factors are intitially stated in a way most favorable to the "multiple selves" thesis; as will be discussed subsequently, the actual discoveries of psychoanalytic theory vary significantly from this idealization. Nonetheless, I proceed to state the ideal psychoanalytic case for multiple selves in order to understand what in the theory and in the phenomena that generated the theory at least suggests such a thesis.

Very generally, three discoveries of Freud suggest that there is more than one person within us all, first, that we have conflicting mental states, and experience them as conflicting; second, that mental states can be given a functional organization; and third, that part of our mental life is unconscious. I shall pursue each of these in the ensuing sections.

3.1. *Discovery of Conflict in a Person's Mental States*

3.1.1. *The Discovery of Simple Conflict*

One should start where Freud himself began, with the fundamental insight that people have conflicting mental states. It is useful to begin with an example from outside of psychoanalytic theory so as to not overload the mere fact of conflict with premature theoretical baggage. Consider the recognizable portrait of conflict drawn by Thomas Schelling (1980), a game theoretician and an economist interest in the rational consumer:

> People behave sometimes as if they had two selves, one who wants clean lungs and long life and another who adores tobacco or one who wants a lean body and another who wants dessert, or one who yearns to improve himself by reading *The Public Interest* and another who would rather watch an old movie on television. The two are in continual contest for control . . . .
>
> How should we conceptualize this rational consumer whom all of us know and who some of us are, who in self disgust grinds his cigarettes down the disposal swearing this time he means never again to risk orphaning his children with lung cancer and is on the street three hours later looking for a store that is still open to buy some cigarettes; who eats a high calorie lunch knowing that he will regret it, does regret it, cannot understand how he lost control, resolves to compensate with a low calorie dinner, eats a high calorie dinner knowing he will regret it, and does regret it; who sits glued to the TV knowing that again tomorrow he'll wake early in a cold sweat unprepared for that morning meeting on which so much of his career depends; who spoils a trip to Disneyland by losing his temper when his children do what he knew they were going to do when he resolved not to lose his temper when they did it?

The temptation to regard ourselves as consisting of several selves starts, as Schelling recognizes, with this basic fact that each of us at least sometimes possesses mental states that are in conflict. We may have inconsistent beliefs, conflicting wants, opposing emotions.

It is a surprisingly difficult task to give an adequate philosophical characterization of this "simple fact" of conflict. Given *de dicto* and *de re* ambiguities regarding the contents of mental states, such conflict is not even clear with respect to beliefs, for which the notion of logical contradiction seems most readily available. And the matter is even muddier with respect to desires and emotions, conflict between which does not seem so amenable to characterization in terms of the logical contradiction between the contents of such mental states.

In any case, assuming that some philosophically respectable account can be given of conflict, one needs to say how second agent temptations begin with this not-so-simple fact. Surely some kinds of conflicts do *not* tempt us in the way Schelling suggests. The world being as it is, we will often, for example, have general desires that conflict in particular circumstances – one cannot both satisfy a general desire to relax and a desire to do well at one's job when the means available to satisfying the one (watching TV) will prevent the satisfaction of the other (being prepared the next day). Such conflict, inevitable simply by being creatures with more than one general desire, should not generate any second agent temptations.

The kind of conflict of desires (ignoring emotions and beliefs) that do generate such temptations are those unresolved conflicts between desires that will *necessarily* conflict on individual occasions. Given the causation of obesity by desserts and lung cancer by smoking, a person who maintains each of Schelling's conflicting desires – long life versus smoking, lean body versus desserts – may seem somewhat irrational in his failure to combine his desires into a consistent preference-order. Such failure is irrational because it is unnecessarily frustrating to act on such unresolved conflicts.[12]

The way in which this leads to second agent temptations is by our assumption of rationality for persons. If we expect a rational person to order his mental states into consistent preference orders, belief systems, and emotional responses, then unresolved conflict itself will lead one to at least think of separate selves, each of whom is more rational (i.e., consistent) than is the whole person with his conflicts. Still, isolated conflicts between particular mental states is only the beginning of this temptation; other discoveries about such conflict must be made before any temptation to talk of multiple selves is very strong.

3.1.2. *Discovery that Certain Types of Conflicts are Recurring*

One way to organize this potential chaos of conflicting states is by aggregating those states by virtue of their instantiating recurring *types* of conflicts. If, for example, the desires to smoke, for dessert, or to watch TV typically conflict not with each other but with the desires to prolong life, promote a lean body, and spiritual improvement, respectively, these might be organized into two groups of conflicting desires. This is, indeed, one of Freud's principal concerns in his definition of ego, id, and superego: one assigns mental states to these structures on the principle of best exhibiting the conflicts that recur within each person (see Arlow *et al.*, 1964.) Ego, id, and superego are thus not evidence of mental conflict, but are themselves aggregations of mental states constructed in a way so as to best exhibit the recurring types of conflict already discovered to exist.

The discovery that mental states do not conflict in a random manner increases the temptation to talk of multiple selves, because it betokens an enduring nature to mental conflict. The temptation to think of "structures" is more natural if there is a consistent pattern of the mental states in conflict to be assigned to such structures.

3.1.3. *Discovery that the Types of Conflicting Mental States Instantiate Intelligible Characters*

Suppose now that we discover that the conflicting mental states aggregated by their instantiation of types of recurring conflicts are just like people we know. For example, suppose our previous principle of organization aggregates the mental states of Schelling's not so rational consumer by putting the desire for dessert, tobacco, and TV, in one system, and the desire for a lean body, long life, and spiritual improvement in another system. These, suitably enriched, might instantiate patterns of character that are intelligible to us as a character that a person in our culture might hold. We might think of gluttons on the one hand and ascetics on the other.

It is important to see that not every pattern of mental states, nor even most such patterns, will instantiate characters that are for us intelligible. Only certain patterns of mental states will be recognizable to us as a potential character structure of a person. These will satisfy two requirements; first, the patterns will have exceeded some threshold of coherence that we expect of any person; and second, these patterns will have within them desires that are intelligible to us as reasons for action, beliefs that are rational in light of the available evidence, and emotions that are appropriate in kind and in intensity to their objects.[13]

Because of these cultural limitations on types of intelligible characters, it is a distinct organizing principle to aggregate mental states by their instantiation of intelligible characters. There is no guarantee that mental states aggregated by types of recurring conflict will be like the intelligible characters of whole persons. Nonetheless, if one were to believe this with respect to the structural metapsychology, one would then characterize the ego, id, and superego somewhat in the way quoted by Ronald de Sousa (1976): "Psychoanalytic theory suggests that man is essentially a battlefield, he is a dark celler in which a maiden aunt and a sex-crazed monkey are locked in mortal combat, the affair being refereed by a rather nervous bank clerk."

3.1.4. *Discovery that One Set of the Recurring Types of Mental States in Conflict is Experienced as Alien*

Loaded into Schelling's portrait is yet another factor: in some sense we might say that Schelling's individual *most wants* to remain healthy, long lived, self-improving and patient with his children. He is not decieved about these desires; he might well say, without deception, that he is conflicted but that he most wants the objects of the "noble" set. Nonetheless, he frustrates this set of desires, manifesting what philosophers since Aristotle might call akrasia, or weakness of will.

For present purposes it is not so important to dissolve the air of paradox surrounding the idea that a person with eyes open can fail to do what he most wants to do. The important point here is how the akratic regards some of his own desires, namely, the ones to which he yields. He may typically regard them as less a part of himself than the "higher" desires that oppose them. If so, this aspect of akrasia, or weakness of will, is but a special case of that general phenomena that formed one of the important bases for Freud's concept of the id. As Morris Eagle (1981) points out in his paper in this volume, some mental states are experienced as 'ego-alien', as not belonging to *me* but to an 'it'. In such cases, one experiences one set of conflicting mental states as outside of the self. If these ego-alien mental states instantiate one of the types of states that regularly conflict, and these ego-alien states themselves form a pattern intelligible as the character of a person, second agent temptations become considerably strengthened. This will particularly be so in the special case of akrasia, because the actions one dislikes can be attributed to 'someone else'.

3.2. *Organization of Mental States by the Functions They Serve*

A second major organizing principle with which Freud was much concerned is the aggregation of mental states in terms of the functions served by such states. Thus, for example, the ego is assigned the function of self-preservation and the functions governing motor control, while the id is assigned those mental functions representative of the sexual instinct. Although Freud (as examined shortly) believed that mental states sorted by this organizing principle would be the same as those sorted by the previous principle of best exhibiting types of recurring conflict, the principles themselves are quite distinct, as can be seen from an analysis of how functions might be assigned to mental states.

Freud and his followers, together with many contemporary philosophers of mind and researchers in artificial intelligence, analogize the mind to the physical body in the sense that each is given a functional organization. What this means for physical medicine is that one assumes certain end states (homeostatic balances) toward which a healthy body tends despite disturbing conditions; with regard to the mental states that are the subject of psychiatry, one likewise assumes some end state of (mental) health toward which various mental states contribute. One in each case indexes a great deal of information about the causal contribution mental or physical states make to the maintenance of the general end state of health, by assigning functions to such physical parts or mental states. That the function of the heart is to circulate the blood is just to say that one of the consequences of the heart's beating is that the blood circulates, and that this consequence itself contributes to the maintenance of that end state with which physical medicine is concerned, physical health. Analogously, to assign to the ego self-preservation as one of its functions is to say that the mental states designated as being part of the ego serve the function of preserving the organism, itself a state which contributes in an obvious way to the overall end state of health of the organism.

It is important to note that when one aggregates mental states by the functional organization attributed to a person, one assigns them to such functionally defined components universally, that is, to all human beings. Such assignment of mental states to various aggregate functions is not (for Freud at least) peculiar to each person; rather, Freud makes the assumption of the existence of universal tendencies, and thus the functional organization of a person should be the same despite differing characteristics of that person's mental states.

There is very little in a functional organization of the mind that should by itself incline one to a multiple self thesis. If psychologists study perception, for example, and subdivide the perceiving that a person does into discrete, functionally defined subroutines that take place when the person perceives, it is the crudest anthropormorphism to posit separate persons as the performers of each subroutine. Rather, a functionalist about minds, such as Dennett (1969), will talk of the "subpersonal" level of description explicitly to avoid the suggestion that his functional subdivisions are to be confused with a *person's* actions or mental states. That there are various stages of information preprocessing going on in visual perceptions, for example, is not to say that the person (or any little person) is doing any of the things described at the subpersonal level.

It is only because Freud thought that a functional organization of the mind would aggregate mental states in the same way as do the conflict principles, that second agent temptations come into being. For if one set of typically conflicting, intelligibly characterized yet ego-alien mental states are just those mental states serving certain universal functions but not others, one may be tempted to think that such set of mental states is more than just a set but is itself a "structure" in the mind (and, with the hoped for function/structure correlation, in the brain as well).

3.3. *Discovery of the Unconscious*

"Unconscious" is ambiguous; it can mean that a person is unconscious, in the sense of not being awake; or it can mean that, although the person is awake, he is unaware of certain of his own mental states. As Morris Eagle (1981) points out, Freud exploited both of the senses of the word, emphasizing the first in his earlier theorizing and the second in his later statement of the theory. I have accordingly separated the discovery of the unconscious into two principles of aggregation: (1) by the characteristic of a mental state being unconscious; and (2) by the characteristic of a mental state being experienced or acted upon while the agent is unconscious.

3.3.1. *Unconscious Mental States*

3.3.1.1. *Discovery that Some Aggregates of Conflicting Mental States are Unconscious.* A basic organizing principle for Freud was based upon the discovery that some mental states are conscious and others unconscious. This was the organizing principle most relied on by Freud in the topographical metapsychology of the "systems" conscious-preconscious and unconscious that preceded (and then uncomfortably coexisted with) the structural metapsychology of ego, superego, and id. One might discover, to revert once again to Schelling's example, that the desire for dessert, for tobacco, and to watch TV are all unconscious desires. If conflicting mental states were sorted by all three principles into the same aggregation, the suggestion of such separate selves, one of whom is conscious and the other of whom is unconscious, becomes that much stronger.

3.3.1.2. *Discovery that Some Unconscious Mental States Appear to Serve as the Rational Motivation for Behavior.* Suppose now that the unconscious mental states that one has discovered not only exist, but sometimes "win" over their conscious opposites. That is, they win in the sense that such unconscious mental states causally influence behavior. Moreover, they influence that behavior in the particular way which delineates a fundamental sense of rationality. An action is rational in this fundamental sense if it fulfills the object of the desire that is given to explain it. For example, one may unconsciously wish to be an artist, and this could cause one to do almost anything, such as cut off an ear or whatever. Even if the existence of such a mental state and its causal connection to behavior were substantiated, however, this would not explain the action as rational because the action cannot be seen as a means to fulfillment of the object of that desire. Freud's discovery that the unconscious was causally active has not been presented as just the discovery that sometimes unconscious mental states influence behavior; Freud also thought that he had discovered that 'the unconscious' acts in the way in which rational agents typically act, adopting means to the ends which they desire.[14] Of course, in most of Freud's examples, some symbolic transformation must take place before one can even entertain the hypothesis that the dreams, parapraxes, or neurotic symptoms of Freud's patients are means to the attainment of unconscious ends. For example, one might find oneself sucking a pen rather than eating the dessert which one truly desires; or one might call out for just deserts as the sublimated form of one's unconscious desire for dessert. While some of these claims are more tenuous than others, a basic claim of Freud is that sometimes the unconscious mental states look for all the world like the use of an action to achieve, in a somewhat attenuated way, the object of an unconscious desire.

The way in which second agent temptations are increased by this discovery should be obvious. First, that one set of the conflicting mental states – say, the more gluttonous desires of Schelling's conflicted individual – causes behavior tempts one to posit an agency with causal powers. Since these states are unconscious, the personal agency we all possess may seem inappropriate; hence, the second agent. Second, if one perceives a pattern of behavior that adjusts itself to serve just those ends that are unconscious, one will be tempted to attribute that behavior to a personlike agent, because the behavior and the mental states that explain it seem to possess an essentially human characteristic, namely, rationality. Morris Eagle recognizes this in his paper in this volume:

> [T]his formulation complicates and deepens the challenge to concepts of unity of the self because in this account all aspects of the complex story are purposive in nature and are occurring simultaneously within the same person. According to the Freudian view, occurring simultaneously are purposive mental maneuvers we normally believe can only

be consciously carried out by a person – including strivings for gratification, avoidance of displeasure, and reaching of compromises – which are now claimed to be carried out unconsciously and are attributed to partial components constituent of the person.

One in short personifies such partial components by attributing to them causal agency and practical reason.

3.3.1.3. *Discovery of Self-Deception.* Self-deception is often pointed to as one of the crucial facts suggesting a multiplicity of selves within any body. Morris Eagle, for example, relies upon an analysis of self-deception (in terms of Fingarette's (1969) idea of disavowals) as "the decisive aspect which threatens unity of personality." Essentially the idea of self-deception amounts to the discovery that there are unconscious mental states that are causally active and rational, only twice over. That is, when one is self-deceived not only does one have unconscious mental states that causally influence action in a seemingly rational way, but there are second-order, unconscious mental states that are also causally active and rational, and what *they* cause is the first-order unconscious mental states to be unconscious. One may, for example, not only unconsciously desire dessert, but one may perform certain manuevers just to prevent oneself from knowing or learning that one has the unconscious desire for dessert.

Self-deception seems to increase the temptation to talk of multiple selves because of its notion that there is both someone who is deceived, and that there is someone who is the deceiver. Deception is a lie, not a mistake, and the difference is that the person who lies knows that what he is saying or implying is false. One who is deceived, on the other hand, is someone who does not know what he is deceived about – else he would not be deceived. Thus, the temptation: one way of reconciling the person who knows and yet doesn't know is to say that there are two persons, one of whom knows and the other of whom is ignorant.

3.3.1.4. *Discovery of the Repressed Nature of the Unconscious.* Suppose now that one gives the explanation for why an individual's desires may be unconscious, and for why he may be deceiving himself about them. Suppose the explanation is itself in terms of an activity engaged in by "someone," who does what he does for an understandable reason. Schelling's consumer, for example, may have *repressed* certain of his desires because of their connection to painful memories from his childhood. Moreover, he may continue to repress them for the same reasons. Since he doesn't view himself as "instituting repression', yet since it too appears to be an activity engaged in for reasons, the temptation is to talk of 'other agents' doing so. Remarkable in Freud's corpus is the number of such repressing second agents, variously termed 'censors', 'gatekeepers', or 'part of the ego'. Such second agents may seem required here in part for reasons already explained in connection with self-deception and the simpler cases of unconscious mental states; repression adds, however, one additional factor. Repression is unlike neurotic symptoms or the slips of the tongue and the like that formed most of the classic evidence for the unconscious, in that "instituting repression" is not a recognized kind of action that persons perform. Repression is much like the processes Freud postulates as part of the dream-work in that none of these inner processes seems to be within the act-repertoire of a person, no matter how intensive the psychoanalysis is nor how frequent his visits to a biofeedback laboratory to increase that repertoire.[15] Because of this, 'little agents' seem necessary if one is to make sense of the talk of *actions* by someone in such cases.

3.3.2. *Discovery of Mental States and Behavior Engaged in While Unconscious*

Long before Freud one might speak not only of unconscious mental states, but also of persons being unconscious in the sense of not being fully awake. A distinct Freudian claim has to do with the explanation of behavior engaged in by the person when he is unconscious, not just when the person is conscious but his behavior is influenced by unconscious mental states. The claim is that the mental states which are either expressed in behavior while unconscious, or that influence dreams while asleep, are just those (unconscious) mental states that explain waking behavior. To revert to Schelling's example once again, suppose that when the individual is dazed by a blow to the head, is rendered unconscious by the nervous shock of being shot in the stomach, is put into a hypnotic trance, or is simply asleep, then out pops the very same mental states thought to exist as unconscious mental states when the person is awake. Thus, for example, he either when asleep engages in somnambulism and walks to the refrigerator, takes out the dessert, eats it, and lights up while he turns on the TV, or if he doesn't actually engage in these behaviors, he nonetheless dreams of doing them. In such a case, the temptation to posit a second intelligent self within the self-same body may increase, first, because there is a sense in which we certainly say that *we* (referring to a self) are asleep or otherwise unconscious. Yet surely *someone* is doing all these things. And second, such repeated appearance of one set of mental states when we are in an altered state of consciousness, may also make us think that there is more conflict than we had thought. It may, that is, lead one to Freud's early views that such unconscious desires are always present in us, waiting for expression whenever we relax our state of vigilance (by being unconscious or in an altered state of consciousness). This is the 'loaded spring' or oozing notion of the unconscious:

> He that has eyes to see and ears to hear may convince himself that no mortal can keep a secret. If his eyes are silent, he chatters with his finger-tips; betrayal oozes out of him at every pore." (Freud 1905)

3.4. *More Pathological Conditions*

The previous three factors exhaust what I take to be those aspects of psychoanalytic theory, and of the data on which the theory was construed, most strongly suggesting separate selves within one and the same human being. Certain particular pathological conditions, namely, multiple personalities, and some of the recent brain research on split brains, may even more strongly suggest reconceptualizing a person as separate selves. Such data are included here because it is sometimes referred to as buttressing the Freudian subdivisions of self, even though it does not form part of the general data for Freud's topographical or structural metapsychology. It is important to see at the outset that such phenomena could not constitute strong support for the *universal* subdivisions Freud thought to exist in all of us, because all of us do not have multiple personalities or split brains.

3.4.1. *Discovery that Behavior Over Time Manifests Several Different Intelligible Characters*

Suppose we alter Schelling's not so rational consumer with the following additional facts: not only are his mental states in systematic, intelligibly characterizable conflict, organized along certain functional principles, and aggregated according to their being conscious or unconscious, but also the person's behavior *over time* matches the aggregations of mental states previously achieved by these three principles. That is, suppose the gluttonous person 'takes charge' for a period of time, and then is supplanted by the ascetic or spiritual individual, who also reigns for time. If one suspends the agency attributions, what this comes to is the claim that the conflicting mental states previously organized by their instantiation in intelligible characters will also find expression in aggregates of behavior that are temporally continuous. It looks like the person is more than one person because he acts so "in character" for periods of time, yet the character he is in changes abruptly from period to period. If one adds the fact of partial memory loss over time, we have an instance of that phenomenon known as multiple personalities.[16]

3.4.2. *Discovery that Each Such "Character" Has Physical Location in the Brain*

Now suppose as a further fact that we discover that the gluttonous 'character' previously discovered to be a character apparently taking charge of aggregates of behavior, also has physical location in the brain. One might think of the glutton being in the right hemisphere, and the more spiritual character in the left. To the extent that our intuitions of what it is to be a person are in part based on being an embodied person, this discovery of the separate autonomous intelligible and intelligent functioning of the two hemispheres of the brain may seem to increase the temptation to speak of separate persons.

If one goes beyond contemporary split brain research to posit the possibility of separating out all physical functions that feed into one hemisphere from those that feed into another, so that there are two separately embodied hemispheres of the brain, the temptation to speak of persons seems to be increased. For now not only is there physical location of the mental states previously characterized in separate regions of the same brain, but there is not even a physical unity of a body which those two hemispheres at least share. (If one is bothered by such recherché philosophical thought experiments, one might think of such splinter bodies as the limiting case of Siamese twins.)

The last factor that might be eliminated is the developmental unity that artificially created splinter bodies would have, that is, they at least *have been* formed as one physical unit. If one were to discover that such splinter persons were a naturally occurring phenomenon, the temptation to regard them as separate persons, given the lack of any developmental unity, would be overwhelming. The limiting case of such splinter persons is, of course, simply two persons.

4. THE PSYCHOANALYTIC CHALLENGE TO THE UNITY OF THE SELF RECONSIDERED

4.1. *The Unity of the Self – I: One Person Per Human Being*

Each of these factors may increase our temptation to speak of a disunified self. Before reassessing this data to judge whether this temptation is warranted, one needs to be clear about what one means by "the unity of the self." The first and most obvious sense of "the unity of the self" is one utilizing the notion of numerical identity of a person. (This I shall henceforth call the question of personal identity, following relatively standard philosophical usage.) This most basic sense of the unity of the self may be defined as the doctrine that there is one person per body, and disunity, accordingly, as there being two or more persons per body. Such a view of the unity of the self need take no position on the metaphysics of the mind/body relationship, and explicitly is not committed to any form of identity theory that urges that a person is identical with his or her body. Rather, the claim of the unity of the self in this sense is only that persons are individuated by their bodies, even though not necessarily identical to them.

A natural contrast to talking of the numerical identity of persons is to talk of the qualitative identity of persons. There is only a limited range of phenomena for which such talk could make sense, if one believed in the unity of the self in the first sense just identified, namely that there is one but only one numerically distinct person per body; for given such a belief, one could not hold that at any given time there was more than one *kind* of person in that body. (One could not hold this because Leibniz's law holds that if two nominally distinct things are in reality one and the same thing, then anything predicable of the one must be predicable of the other, and vice versa.) Accordingly, at any one time, if there is unity of the self in the sense of numerical identity, there must also be unity of the self in the sense of qualitative identity.

Where this will not be true is for the identity of persons over time. In this context, we often do say of a (numerically distinct) person that he is not the "same person" as he was at some earlier point in time. What we mean when we say this is not that he is not the same person that we knew before; rather, such language is simply a dramatic form of saying that he is not the same *kind* of person as he was.

The unity of the self which this notion of qualitative identity would define, would simply be the doctrine that over time persons remain the same kind of person as they were previously, and disunity, the denial of this claim. There is nothing very problematic about this sense of the unity of the self, nor is any disunity discovered in this sense very troublesome to the basic assumptions of metaphysics or morals earlier discussed. Hence, it is worth distinguishing this other sense of personal identity only to put it aside from further consideration.

If psychoanalytic theory could make out the idealized case recreated around Schelling's consumer, buttressed perhaps by the findings of multiple personality or split brain research, would it show us a disunity of the self in the sense just set forth, *viz*, more than one person per body? My inclination is to say that it could. If one discovered that certain types of mental states are regularly present and in recurring conflict with one another; that each of these aggregates of mental states in conflict were coherent enough and intelligible enough to constitute characters of a person; that the mental states composing at least one of these aggregates were experienced as belonging to another (if they were conscious at all); that these aggregates of mental states are just the aggregations one would achieve if one sorted such states by the functions they served; that these aggregations of mental states are also just those aggregations one would achieve if one sorted such states by their being conscious or unconscious, or by their being manifested when the agent is conscious or unconscious; that the unconscious mental states included full practical syllogisms causing actions by some agent; that such actions included deceiving one's consciousness; and that such actions included activities, such as repression, that no consciousness is aware of or becomes, even after psychoanalysis, aware of ... if one discovered all of this, I suspect that we might embrace Freud's (1924) conclusion that "we are not masters in our own house" but that 'someone else' – id, unconscious – is.

One could only decide this question by having ready to hand some successful analysis of personal identity that could be defended against the extended thought experiments in the philosophy of personal identity. Since such thought experiments seem to present counterexamples to taking any of the three leading candidates of personal identity – spatio-temporal continuity, consciousness and memory, or coherence and consistency of character structure – as the criterion of personal identity,[17] I shall take a different tack, which will be to re-examine the Freudian case.

The presentation of the Freudian case for disunity has been idealized. It is now time to ask whether Freud actually made out all or even most of his idealized case for disunity of the self. To begin with, there is good reason to believe that Freud's three major principles for aggregating mental states will not in fact sort them into the same sets. Freud himself (1923) conceded part of this point when he shifted from the topographic metapsychology to the structural; "part of" the ego was unconscious, he thought, namely that part that does the repressing. One accordingly cannot assume that mental states functionally assigned to the ego are conscious (nor, one might add, is there any reason to believe that all mental states having to do with sex are unconscious). Likewise, the conscious/unconscious principle does not seem to map onto the conflict principle. Many conflicts, even of a recurring nature, are between mental states both of which are quite conscious; analogously (at least if one believes many parts of the Freudian theory) many conflicts of a recurring nature are between unconscious mental states. Conflict is not necessarily between mental states allocated to different topographical systems.

Less often noticed is the necessary lack of fit between the conflict principle and the functional principle. If one aggregates mental states because of a functional organization of the mind, one will do so universally, for all persons. Yet conflicts surely vary greatly for different people. The conflicting emotions Dora felt for the gentleman who propositioned her (Freud, 1907) seem to have little resemblance to the conflicting intentions Freud used to explain slips of the tongue (Freud, 1900). Only by relying on an extraordinarily reductionist notion about mental states – such as an instinct theory that reduces them to either sex or aggression – can one make such differing conflicts in fact be of a more general, universal type. Without such a (highly suspect) reduction, the mental states sorted by the highly individualized conflict principle will not be sorted into the same sets as is done by a function-based assignment.

Even within the general aggregating principles considered by themselves, the differing strands of each of those principles will not sort mental states into the same sets. With regard to the unconscious subprinciples: only with some extraordinary transformation can one discover that dreams are the 'royal road' into those unconscious mental states that explain conscious behavior. The mental states people unproblematically have when they dream (the manifest content) are not the same as the mental states Freud uses to explain neurotic symptoms. Similarly, the mental states one experiences or acts upon in hypnosis may include some of those unconscious mental states on which one acts when fully awake, but seemingly many of the former states are not what would be called 'unconscious' if possessed when awake.

With regard to the subprinciples of conflict: there is no reason to believe that the mental states aggregated upon the subprinciples of best exhibiting recurring types of conflicts and of being experienced as ego-alien, on the one hand, will instantiate intelligible character structures; indeed, the parody given of the superego, id, and ego as the maiden aunt, the sex-crazed monkey, and the nervous bank clerk, betrays the fact that these are not intelligible characters in the way that multiple personalitied persons have different, intelligible characters within them. Rather, these are caricatures. One can see this by pressing the kinds of questions Irving Thalberg has asked regarding Freud's mini-agents. Consider, Thalberg advises, the ego's wish to sleep, one of the two main motives for dreaming of Freudian theory:

> For *whose* sleep does it yearn? Mine? Just its own? Who is the owner of the "free attention" which is on duty? Whose waking may the attentive "guard" consider 'more advisable' that a continuation of sleep? His own? Mine? When we hear of the sleeping 'townsman', we should raise further questions. On his overall view of human motivation, can Freud suppose that any 'watchman' is "conscientious" enough to care about his fellow townsmen? Does he call them from sleep just to help him put down rowdy instincts? What harm can the most licentious instincts do him? If Freud were talking about ordinary sentinals and sleeping townsmen, we would be able to answer. (Thalberg, 1973, at p. 161.)

We would be able to answer because whole people have characters that allow us to decide whether, e.g., an altruistic response is in character or not. Ego, id, or superego – or their subagents of censors, watchmen, etc. – are insufficiently rich in character to answer such questions.

Aside from the lack of fit between the different organizing principles and subprinciples, there are more serious problems with the idealized case for disunity of self. To begin with the fundamental fact of conflict: striking in reading psychoanalytic theory is how often conflict is *posited* for reasons of theoretical neatness, rather than being a datum that is discovered and need be accounted for by the theory. An example is Freud's theory of dreams (where conflict does not naturally suggest itself as it does in some parapraxes and many neurotic symptoms): Freud simply posits a wish to sleep conflicting with the unconscious wish from childhood seeking expression in a dream. Freud does so because he wished to maintain the parallel he thought ought to exist between dreams and neurotic symptoms: both, he thought, should be viewed as compromise formations between mental states belonging to different topographical systems. While one undoubted fact of mental life is that there are conflicts, assessing how much conflict there is, and whether it fits a certain pattern, is not aided by a theory that posits just those mental states necessary for there to be conflict of the right pattern.

With regard to the discoveries about the unconscious, there are several points to be made. The first is a quibble about the amount of our mental life that is unconscious. The Freud who liked mechanical models – of an electrical, chemical, or hydraulic nature – believed that all conscious mental states are underlain by unconscious ones. This "iceberg" conception of the unconscious (wherein consciousness is but the tip above water) is part of the economic metapsychology (psychic 'energy') justifiably written off by some influential theoreticians about psychoanalysis (Schafer, 1976; Nagel, 1959). Moreover, this iceberg conception obscures a more legitimate conception of "the unconscious" in Freud's work, namely, as the name for those mental states not presently known to their possessor but recapturable by extended memory (Moore, 1980a). If one has in mind the latter notion of unconscious, unconscious mental states do not appear everywhere; they are (perhaps) manifested in dreams and some behavior, and in principle are recapturable by extended memory (including re-experiencing during transference).

Second, there is in psychoanalytic theory an enormous extension of action language to behavior caused by unconscious mental states. This is true of Freud's own theorizing, but also of the recent reconceptualization of the theory done by Roy Schafer (1976). Such Freudians assume that if one discovers an unconscious desire, for example, that causes a dream, therefore dreaming is an action. As I have argued at length elsewhere (Moore, 1980a, 1980b), this connection needn't hold. Dreams could be wishfulfillments in that their contents fantasize situations in which wishes the dreamer has are fulfilled, and dreams could be caused by such wishes; even so, they need not be actions, as the "wayward causal chain" counterexamples in the recent philosophy of action illustrate.[18] The less of dreams, parapraxes, or symptoms that are conceptualized as human actions, the less of course is the need to find an agent whose actions they are.

Third, with regard to repression and the other mechanisms of defense, one should systematically reconceptualize Freud's mini-agent stories from purported tales of actions for reasons into functionally characterized subroutines requiring no personal agents, however small. Repression, for example, not being an action persons engage in (nor even recapture with extended memory that they have engaged in), should be seen as the process that must take place if some mental states are significantly harder to recapture than others (that is, are unconscious rather than preconscious). Likewise, stories about little agents engaged in displacement, condensation, pictorial representation, and secondary revision (the four categories of the dream-work that distort the manifest content of the dream) are better retold as functionally characterized subroutines that must go on if dreaming is caused by unconscious wishes that cannot be given direct expression. Or as a last example, consider the ego's (or the preconscious') wish to sleep: nothing essential to Freud's guardian of sleep hypothesis is lost if this is not a wish assigned to an *agency*, but rather, is thought of as a sleep-preserving *function* of the *process* of dreaming.[19]

Self-deception may not be as easily dealt with as this, for it and certain other phenomena, such as resistance to therapy, may be genuine instances of unconscious actions. That is, one may well not be able to reconceptualize such phenomena as either processes serving functions or as (non-action) behavior caused by wishes, for either of such reconstruals seems to leave out an important part of the Freudian claim: the person is acting to deceive oneself, or to resist treatment, and will (in principle) come to see non-inferentially that these were indeed actions by him.

Granting that all of this may be true about self-deception and resistance does not strengthen the case for disunity of the self. Rather, the opposite would seem to be the effect of so regarding these phenomena. The (single) person deceives himself and unconsciously resists a recovery he himself desires. The semantic puzzle about self-deception mentioned earlier – the apparent contradiction of knowing and not knowing – can be dealt with in ways less costly than positing two persons. One might, for example, vary the senses of "know" such that a (single) person may without contradiction be said to know something to be the case and yet not know it. He may know that some proposition is false, for example, in the sense that he can recapture with his extended memory such a belief and find it reflected in his behavior, and he may not know that that proposition is false in the sense that he was not aware (conscious) of it. (See Moore (1980a).)

In these ways, one can grant Freud's insights about the unconscious without giving rise to any second agent temptations. What of the more recent phenomena of split brains or multiple personalities? While the multiple personality cases require more extended discussion, which I shall pursue in the succeeding section, one can be more abbreviated with the phenomena of split brains. Nothing about such phenomena should convince us of there being two or more persons per human body; after all, the fact that there are anatomically isolatable centers of autonomous, intelligent functioning should come as no surprise. Presumably any information processing system will have correlations between the functional subroutines necessary for the system to process information intelligently, and the structural characteristics of that system. If we had not discovered such function/structure correlations between crude anatomy and mental functions, we should surely have discovered such correlations in more complex and sophisticated electro-chemical terms. There is thus, ultimately, nothing remarkable or particularly challenging about the fact that the left hemisphere and the right hemisphere of the brain can function relatively autonomous of one another once the connecting tissue has been severed. Such could be true of any functionally defined 'centers', no matter what their structural characteristics turned out to be.

I conclude that psychoanalysis has not made its case for disunity of the self in the fundamental sense of more than one person per body. Although Freud clearly and continuously talks in a way that seems to presuppose disunity in this most fundamental sense, none of the data on which he has relied in fact justifies such talk. To talk of ego, id, superego, censors, watchmen, guardians, systems unconscious, and the like, performing actions for reasons, intentionally, with their own mental states, awareness, character structure, and responsibility, is simply an anthropomorphic mistake. Nothing would be gained by personalizing these functional subdivisions of the mind, and a great deal, in terms of our moral, legal and metaphysical needs, would be lost.

4.2. *The Unity of the Self II: One Personality Per Person*

A second sense of "the unity of the self" distinct from the fundamental sense just explored is found by ceasing to speak of persons and move to speaking of personalities in the sense of character structures. There may be problems about speaking of the *identity* of personalities, because of the problems in regarding personalities (or characters) as particulars. Personalities are particulars only in what Bernard Williams (1973) calls a "weak sense," as can be seen by the fact that for personalities there is no real separation of qualitative from numerical identity. If, *within the same person*, two personalities are qualitatively identical, then they will also be numerically identical. (Thus, for the personalities of any given person, the other part of Leibniz's thought, commonly called the identity of indiscernibles, will be true, even if it is not generally true.) Because of this, one may want to think of personalities as universals so that talk of different characters within one person is just talk about that person and his (differing) mental states and behavior.

In any case, if one can regard personalities as particulars, the second sense of unity of self will be defined as one personality (intelligible character structure) per person, and disunity as two or more personalities per person. Unity will not presuppose that a person is identical with his or her personality, but only that if there is unity of the self in this second sense, there will be at most one personality per person.

It is an interesting question how the phenomena usually referred to as "multiple personalities" should be conceptualized. Surely in some sense the phenomena suggest disunity of self, but the question is, in what sense? There seem to be three possibilities. First, as Morris Eagle observes, such phenomena seem to challenge what he calls "numerical unity" and what I call the unity of self in its most basic sense, that is, one person per body. One might thus think of "multiple-personalitied persons" as really being separate persons, even though acting through the same body. Alternatively, one might utilize the second sense of the unity of the self just defined, and talk of there being more than one personality (as a particular) per person. Third, one might simply talk of such persons as manifesting different characters over time; they are not, that is, the same *kind* of person at different times.

The phenomena of multiple personalities do not seem adequately conceptualized in the third of these ways. For the third, using the idea of qualitative identity of a person over time fails to capture several salient features of multiple personality cases: (1) There seems to be a dramatic difference between normal people who act out of character and multiple personalities who act "out of character" for one self but in character for different "selves." The difference seems to lie in the coherence and the intelligibility of the characters formed by aggregating the mental states 'out of character' for the multiple personalitied person. Such sharp breaks between different, intelligible characters may incline us toward regarding such personalities as particulars and not as universals. (2) Multiple personalitied persons not only possess different characters, they also appear to experience 'themselves' as being different "characters" (now in the literary sense). The data seem to be that only sometimes do the personalities "know" each other; that when they do, they may regard them as separate; that they think of self-interest in terms consistent with the character adopted by that personality, not by some overall self-interest; and that there is a good deal of amnesia with regard to the mental states and behavior taking place when other personalities are in charge, at least with respect to emotion-laden issues (Ludwig *et al.*) All of this suggests a disunity much more radical than that described as "not being the same kind of person" at all times. (3) Lastly, multiple personalitied persons appear to have different characters that are co-conscious and not just appearing at different times.[20] Qualitative identity *over time* cannot capture this disunity *at a time*.

Whether multiple personalities should be thought of as being disunified in the first or senses distinguished earlier is a more difficult question. Certainly some of the basic assumptions we make about persons are challenged by this phenomena: Locke's claim that our experience is of one self, the presupposition of an atomic rather than a molecular self-interest of an egoistic moral theory, and perhaps even the unity of self presupposed by our responsibility assessments, are sufficiently threatened that one might think that here at least there is disunity in its most fundamental sense, namely, more than one person per body. On the other hand, certain of our basic metaphysical and moral assumptions about persons are not shaken by the multiple personality phenomena. There is still but one physical body to be acted through, and but one physical brain to which mental states may be related in some non-dualistic way. Likewise, our assignment of rights by theories of distributive justice or of property, still seems applicable to persons as individuated by bodies. The phenomena of multiple personalities at best suggests a standoff on the unity of the self in its most basic sense; for pitted against the fact of a single physical embodiment are the facts of radically disunified character and discontinuous conscious experience, and pitted against the unity suggested by our assignment of rights to such persons is the disunity suggested by our absolving them of responsibility.

One way of resolving this apparent standoff would be to defend either the spatio-temporal continuity criterion of personhood or the experiential and characterological criteria, and come out accordingly for or against unity of self (in its most basic sense) for multiple personalities. Alternatively, one might view such persons in the same way as we regard others who are mentally ill, as having 'suspended personhood'. There are entities that we recognize are not (fully) persons even though they (the *same* entities) may in the future become persons, namely, young children and the mentally ill. (Moore, 1975, 1980c.) The suspended personhood of such entities is recognised most dramatically in the legal and moral spheres: they are not accorded the full panoply of rights held by sane adults, not held to be proper subjects of responsibility, and not held to be able to calculate their own self-interest. Such suspension of personhood is also reflected in the lack of another basic attribute of being a person, rationality.

Multiple personalitied persons should be regarded as but a special case of suspended personhood. There was but one person originally, and (if therapy is successful) there will be but one person again.[21] During the time that intervenes, there may be no answer to the question, "how many persons"?; but such lack of an answer could be due to the fact that there isn't even one.

In any case, however one comes out on conceptualizing multiple personalities, such phenomena are quite rare and cannot be the basis for the general psychology that psychoanalytic theory purports to be. They are not universal characteristics of persons, but only pathological conditions applicable to a limited class of persons.

The more universal phenomena earlier described as "the data of psychoanalysis" do not show a disunity of self in anything like the sense(s) in which multiple personalitied persons do. They don't show this for two reasons. First, once one leaves the phenomena of multiple personalities, one leaves behind the crucial fact of there being temporally continuous behavior manifesting the different personalities posited to exist within the single person. That is, neither Freud nor anyone else would claim that for the different 'characters' of the id, ego, or superego, for the system unconscious or for their subagents, that there is behavior over time manifesting just those 'characters'. One does not have an "id character" as one who is multiple personalitied might have a gluttonous character. Rather, the "characters" of the id, ego, and superego are manifested in behavior only in the sense that certain bits of behavior may be said to be due to them, but there is no large aggregation of behavior that can be said to be exclusively due to them; rather, aggregations of behavior are to be explained as being due to a mix of all three such 'characters'.

The importance of this difference is this: without the temporally continuous behavior manifesting these different characters, there will be no unique or correct determination of what sorts of characters one may possess. Psychoanalysis will join other interpretative schemes relying on 'ideal types', and will share with them one of their most serious problems of method, namely, the seeming total lack of any criterion of correctness for selection of the pure ('ideal') types. What, for example, is to prevent a Nietzschian psychologist from positing Dionysian and Apollinian characters, neither of which 'takes charge' for any aggregates of behavior, but each of which influences every piece of behavior in which one engages? Once one is free to explain all bits of behavior as mixes of these character types, there seems to be nothing but esthetics as a ground on which to prefer one interpretive scheme to another. Although one may posit behavior to manifest ego, id, and superego, one may equally well posit it to manifest Dionysian and Apollinian characters, and there seems to be no room for intelligent debate.

The second reason that the data of psychoanalysis do not show a disunity of self, even in the sense of more than one personality per person, has already been mentioned before: ego, id, superego are all very "flat" characters, not the kind of 'round', or well-developed, characters we find in good fiction and real life. True multiple personalities, by way of contrast, are much more like character structures of whole persons in their richness and intelligibility.

4.3. *The Unity of the Self III: One (Integrated, Congruent) Self Per Person*

The foregoing, I think, disposes of any serious threat to the kind of unity of self presupposed by our moral, legal, or metaphysical needs. The Freudian case for multiple persons, or even personalities, is despite its initial appearances, quite weak. Having said all of this, however, there remains the nagging suspicion that one has avoided some sense of "unity of self" such that one can talk of disunified selves. After all, the temptation to talk of disunity is not limited to Freud, as the Schelling article indicates. Indeed, phenomena such as "standing back from oneself" (self-consciousness), or Frankfurt's (1971) second order desires about one's own first order desires, join Schelling's battle for self-control as the kind of common, everyday experiences that suggest separate selves.

The temptation for philosophers discussing personal identity is to regard the unity of the self in one of the above two senses because they employ the familiar notions of identity as a relation between distinct particulars. This allows one to talk about personal identity in a philosophically familiar way. Much of the discussion of personal identity in psychoanalytic theory, or psychology in general, cannot be reduced to a discussion in terms of real identities between real entities (particulars). One can see this because of two characteristics of such talk by psychoanalysts and others: (1) the notion of self-identity employed by psychoanalysts must be about a 'scalar' phenomenon, that is, a matter of degree. The unity of the self is often regarded as an achievable state, something that a person who is successful in psychoanalytic therapy reaches. Saying this about the unity of the self means that one is discussing a scaler phenomenon in the sense that the identity presupposed by the unity of the self is something at which one can be more or less successful. It will be a matter of degree, not the all or nothing kind of question true identities between distinct entities will raise. As Morris Eagle once suggested to me, discussing "unity of the self" in either of the two preceding senses will allow one to talk about one person or several, but there will be no room for "fractions." (2) The self-identity commonly discussed by psychoanalysts is rather clearly tied to psychological experience. The unity of the self discussed is not a relation between entities; rather, it seems to be a *sense* of self-identity that a person possesses, without taking seriously any apparent ontological commitments to real entities.

Each of these characteristics suggests that further senses of "the unity of the self" are necessary if one is to capture what psychoanalysts commonly discuss under that rubric. It is in such senses, I think, that Eagle's 'disunity of self' is to be found, as he recognizes.[22] The first such additional sense of the unity of the self is what I shall call the sense of self-identity, and the second, what I shall call the congruence between the mental states one includes as part of oneself and those that form part of one's total personality.[23] I shall discuss each briefly in turn.

4.3.1. *The Sense of Self-Identity*

This is the sense one has when one has crossed some threshold of the coherence of his or her mental states. One does not, that is, regard oneself as the possessor of a *chaotic* pattern of mental states, but rather regards oneself as having a relatively well-ordered set, a preference order, a relatively consistent set of beliefs, and a relatively non-conflicting set of emotional attachments.

The opposite of someone who is unified in this sense is not someone who experiences multiple selves within himself, but rather is the person who loses his sense of self entirely. If one believes R. D. Laing (1959) on the phenomenology of schizophrenia, schizophrenics often lose just this sense of I, even losing their ability to differentiate themselves from their external environment. Heinz Kohut (1971, 1977) in his recent restructuring of psychoanalytic theory around this idea of the self, tells us that what he calls narcissitic personality disorders and borderline conditions also give rise to this experience of an unintegrated, non-cohesive self.

The unity of the self in its third sense should be defined as the achievement of a certain threshold of coherence of mental states, and disunity of self as a failure at that essential task for any person.

Some of the data of psychoanalysis do indeed show a lack of unity of self in this sense, some persons do have a diminished *sense* of self-identity. However, one should remember that these are a limited class of cases, not a claim across the board to have shown that all of us have disunity in this sense. And second, one should also recall that the disunity that is the opposite of unity is not that there is more than one self in any sense; rather the opposite is that one lacks a sense of (a whole, integrated) self at all. The experience of being more than one self really only comes in the multiple personality cases with memory discontinuities. More typical disavowed or ego-alien mental states are not themselves cohered into a sense of second self; they are simply alienated as being "not me."

For both of these reasons, the lack of a unified sense of oneself, although an important phenomenon for certain pathological conditions, is not and cannot constitute a general challenge to the unity of self presupposed by our moral, legal, and metaphysical needs.

4.3.2. *Congruence of the Sense of Self and the Total Personality*

The last sense of the unity of the self is once again dependent upon the phenomenology of self-identity, that is, upon the sense of self-identity that a person achieves. Here, however, the unity of the self is not to be defined as the coherence or non-conflicting nature of one's mental states. Rather, assume that one has a strong sense of self, that is, the mental states one consciously affirms as part of the self are relatively conflict-free, coherent, etc. Nonetheless, there may be lack of unity of the self in the sense that the mental states thus affirmed as part of the self do not include significant portions of the total personality of the person. Unity of the self in this fourth and distinct sense will thus be defined as a congruence between the mental states and traits of character identified as oneself, and those traits of character and mental states that are parts of one's total personality. Disunity, accordingly, is a lack of congruence between one's actual mental states and those which one affirms as constituting oneself.

If one believes the data of psychoanalytic theory, all of us suffer from some form of this disunity. Indeed, the entire notion of the mechanisms of defense is a charting of the various ways in which there is excluded from our sense of self-identity those aspects of our total personality which are painful, threatening, or in some way unpleasant. One of the most important ways of defending one's sense of self from these threatening mental states is to render them unconscious. This is Freud's notion of repression, a process by virtue of which those mental states of which the person is ashamed, unable to affirm, and the like, are kept from consciousness and thus from one's sense of oneself. Similarly, even if one does not render the mental states unconscious, one may experience them as being ego-alien. That is, certain thoughts may occur to one which one says are not me, not part of myself, but seem to come from someone else. The thoughts are not unconscious, but just disassociated as being part of one's conscious sense of self. A third maneuver ('displacement') is to rob one's own mental states of their true emotional significance; as before, such defensive maneuvers do not render the mental states unconscious, but only make them seem unimportant because the emotional significance which they do possess for the person is not consciously experienced by that person.

If certain mental states are unconscious, or disavowed in some other way, one will have a sense of self divorced from the total personality, and thus, disunity of the self in this fourth sense. Such disunity is true of all of us to some degree, the degree depending upon how well we have integrated the mental states constituting our total personality into our conscious sense of self. But as before, the opposite of unity here is not that there are several selves but only that one has not successfully integrated parts of himself into his sense of self. Only in multiple personalities do we get any second agent temptations, that is, do we have a sense that the mental states or traits of character excluded from the sense of themselves form or are to be regarded as, or experienced as, a second self. There is thus no serious challenge to our basic needs for the unity of the self by a showing of disunity in this last sense.

5. THE DISUNITIES OF SELF AND PSYCHOANALYTIC STRUCTURALISM

To the extent that the data of psychoanalysis, or related matters, shows us that there are significant disunities of self, to what extent are the topographical or structural subdivisions of the Freudian metapsychology an appropriate way to conceptualize those disunities? Freud himself seemed to have assumed that he had shown disunity of the self in its first and most fundamental sense, that is, that there is more than one person per body, given the degree to which he attributed states attributable only to persons to the subagencies of ego, id, and superego. However, as argued above, the data do not support such a claim, and there is thus no point to asking whether such 'disunity' should be conceptualized in these terms, for the simple fact is there is no such disunity to be so conceptualized.

Although there are disunities in the sense put to the side earlier – the sense in which we might say of a person, he is not the same *kind* of person he was the day before – these are not disunities that in any way track into the topographical or structural subdivisions of self. One would have to change the developmental part of psychoanalytic theory considerably before one came to a notion of development such that one was at one stage of his development pure ego, at another time pure id, etc. Since neither Freud nor anyone else makes such a move, nor is it very tempting independently of its history, this sort of disunity being conceptualized as ego, id, superego requires no further discussion.

Disunity in its second sense, as being more than one personality per person, is, as before discussed, largely limited to the situation earlier described as multiple personalities (if there). This phenomenon also does not track into the functionally organized, conflict reproducing distinctions of ego, id, or superego, nor into those distinctions in terms of conscious, preconscious, and unconscious. These are different sets of distinctions because the kinds of personalities a multiple personalitied person has are not the same as the "personality" that the ego, id, superego, or the conscious and unconscious components might be said to have. The fact that this disunity may exist for a limited class of persons is no justification for the metapsychologies in question. Freud never claimed to the contrary, so this sense of disunity also is not to be conceptualized in terms of the topographic or structural metapsychology.

The third sense of disunity of self, having to do with the lack of cohesion of those conscious mental states one affirms as part of one's self, is more arguably to be conceptualized in terms of either the topographical or structural metapsychology. Indeed, it is common for psychoanalysts to discuss this range of phenomena as a "splitting of the ego". One wants to raise three questions about such a way of talking about these phenomena. First of all, there is the simple matter of clarity. We have a comprehensible way of discussing this phenomenon in terms of a lack of a sense of self. One has to ask whether anything is gained by taking a language we already understand and replacing it with a language which we do not antecedently understand, particularly if the new language is simply the provision of a set of synonyms for the old (that is, we talk of ego rather than self, and a splitting of some "thing" rather than a lack of a sense of self-identity). Such multiplication of vocabulary is both less clear because of its seeming ontological commitment and its seeming increased precision, and is, in any event, superfluous.

Second, such a stipulation of a synonym for the self in terms of the ego would not be to justify the *tripartite* division of the self, which would include such other items as superegos and ids. For recall, the disunity of self this sense of disunity encompasses is not a disunity of multiple selves in any sense; rather, someone who has not integrated his conscious mental states into a strong sense of self-identity is someone who has no sense of self, not other selves. Hence, even if one wishes to supply a synonym for the self in terms of the ego, it would be no justification for supplying such a synonym for the remaining entities postulated by the topographical or structural theory.

Third, in order to have the concept of the ego do the kind of theoretical work psychoanalysts want it to do, one will have to give the word at least two senses. One will talk of the ego as a synonym for the self when one uses 'splitting the ego' as a proper description for a lack of a sense of self-identity. On the other hand, when one aggregates mental states by their functional contribution to the overall functioning of the organism, one will seemingly have given a different definition of the 'ego', different in the sense that the mental states includable within it will not be the same as the mental states includable when one talks of it as a synonym for the sense of self-identity. This will be true because of the fact adverted to earlier: the sense of self-identity will vary with each individual; indeed, it would have to, since it is an achieveable state, something at which some people are more successful than others. By way of contrast, the ego defined by the functional organization of persons would universally include a standard set of mental states, and will not be variable between persons. There is no reason to expect, and every reason not to expect, that these two ways of aggregating mental states will aggregate just the same states. A sense of self peculiar to each person will seemingly sometimes match, and as often not match, the ego defined by the functional aggregation of mental states. Simply put, the sense of self-identity is unique to each person, whereas the functional organizations of persons is not.

Some of these same problems will infect a conceptualization of disunity in its fourth sense in terms of the topographical or structural metapsychologies. It will be recalled that the sense of disunity in its fourth sense is the lack of congruence between one's sense of self-identity, and those mental states truly a part of one's total personality. Thus, if the id is to be conceptualized as all those mental states which a person, via the mechanisms of defense, disavows or represses in some way, then one has simply provided a synonym for the disavowed part of the total personality. As before, little seems to be gained by adding a technical-sounding word for what one already understands, namely, that there are mental states that are experienced as not being part of oneself but which are nonetheless part of one's total personality.

Even more fundamentally, the last objection advanced against Freud's conceptualizing of the third sense of disunity applies here as well. The universal, functional definition of the id will simply get in the way of the personally variable parts of the total personality that each person disavows or represses. It simply will not be the case universally, for example, that a person's sexual wishes will either be repressed or disavowed in some other way. Sometimes that will be the case and sometimes it will not. Thus, when 'id' is used as a synonym for that part of the total personality not affirmed as part of oneself, sometimes such sexual wishes will be included in the id, and sometimes they will be excluded. The problem lies in the fact that this highly individualized mode of aggregating the mental states belonging to the id will not match the universal attribution of those sex drives of any person to the id demanded by Freud's functional organization of the person. As is well known, basic to the drives functionally assigned to the id are those of sex. This lack of congruence between the id defined as the excluded total personality and the id defined by its functional organizing principle will result in two senses of "id," just as it will two senses of "ego." The result of such two quite different things being named by the same name is confusion, not theoretical advancement.

6. CONCLUSION

The central theses of this paper are simply put and are two in number: first, that while there are some disunities of self revealed by psychoanalytic theory and by related data, they are not the sort of disunities that raise havoc with our basic metaphysical, moral, or legal presuppositions of who we are. Second, that for what disunities there may be, the Freudian structuralist or topographical metapsychologies are not useful ways in which to conceptualize them.

University of Southern California

NOTES

* Professor of Law, University of Southern California Law Center. The paper was initially presented as a commentary on Morris Eagle's paper at the Third Annual Conference of the International Union for the History and Philosophy of Science, and was later given at a colloquium of the Philosophy Department of Stanford University. My thanks go to the participants at both these presentations (and particularly to Michael Bratman, Morris Eagle, Herbert Morris, and Stephen Morse) for their many helpful comments. Thanks also go to David Vieweg for his research assistance.

[1] See Dennett (1976) for a discussion of these two kinds of needs for some concept of a person.

[2] See de Sousa (1976), who urges against a disunified self account of weakness of will, that mental states would be needlessly multiplied in this way.

[3] An identity theorist might try to hang onto the multiple selves thesis by giving up the individuation of mental states by persons. He might, that is, say that different persons could have the same mental state-token. That, interestingly enough, commits one who believes in multiple selves also to believing in *group minds*.

[4] For a fine-grained approach, see Goldman (1970); for a more coarse-grained view, Davidson (1980). For a comparison of the two approaches, see Brody (1980).

[5] A basic act is an act I do without doing some other act in order to do the basic act. I may, for example, open the door *by* moving my arm, but my moving my arm on that occasion was a basic act if I did nothing else in order to move it. See generally Danto (1965), Goldman (1970), and Pears (1975).

[6] Spelling out this relation is one of the basic questions in both the philosophy of action and the philosophy of law. Since Wittgenstein (1953), it has seemed difficult to urge that the relation is simply one of identity.

[7] A good thought experiment to test this is to ask how many basic acts are performed when Siamese twins move a shared body part. I think our intuitions are that we don't know what to say about such cases, because our individuation by movements suggests that there is one basic act whereas our individuation by persons (or by persons' 'willings' or 'volitions') suggests there is two. The point of the text simply is that multiple selves within the same body will generate such puzzles about *all* basic acts.

[8] For a nice summary of this, see Wiggins (1980).

[9] Locke's property theory is spelled out in Locke (1956). Locke's theory on the acquisition of property rights is not only a political theory; it also finds reflection in numerous doctrines of property law, such as the acquisition of possession rights in land, wild animals, water, copyright, and the like.

[10] There are puzzles enough about responsibility, even without the multiple self thesis, if one recognizes unconscious mental states. See Moore (1980a).

[11] Irving Thalberg (1973) pursues this line of attack against any personification of Freudian subdivisions of the mind.

[12] Cf. Mullane (1971), who argues that Freudian explanation show the irrationality of his patients because of this 'unnecessary' frustration. For a discussion of this sense of 'rationality', and others that are more basic, see Moore (1975).

[13] Each of these matters is explored, respectively, in: Watt (1972), Ackerman (1973), and Sachs (1973).

[14] For an explication of the way in which desires must 'rationalize' an action if they are to be reasons for action, and for an analysis of Freud's attempt to fit his insights about the unconscious into this form, see Moore (1980b).

[15] For an exposition of a basic act-repertoire, see Danto (1968). The four processes of the dream-work were: displacement, condensation, pictorial representation, and secondary revision. For a discussion of each of these processes, see Moore (1980b).

[16] On multiple personalities generally, see Prince (1930) and Ludwig and Bundfeldt (1972).

[17] See Brody (1980) for a summary of the now voluminous debate about these proposed cirteria of personhood.

[18] Such examples are of the following sort: imagine that A wants to run over a certain person with his car; at time t he also believes that if he pushes down his foot on the accelerator, then he will run that person down; his want and belief get him so excited that before *he* can push down the accelerator, his foot slips off the brake pedal onto the accelerator, causing his car to run down the intended victim. Although his slip was *caused* by a belief and a desire fitting the form of a practical syllogism, his slip remains a slip, not a basic act. See generally Moore (1980b).

[19] This last point is argued for at some length in Moore (1980b).

[20] Taylor and Martin (1944) distinguish coconscious personalities existing simultaneously from alternating personalities, which appear only in sequence over time.

[21] Therapy for such persons is usually thought to involve a 'merging' or a "fusion" of the separate personalities into one. Such fusion involves the disappearance of the memory blockage and the character discontinuities. Ludwig, Brandsma, Wilbur, Bendfeldt, and Jameson (1972). Insofar as such fusion involves recapture of the memories of what one felt and did while in different personalities, as Morton Prince (1906) claimed in his classic study, one could even urge that there was one and the same person there all of the time. For then there would not only be spatio-temporal continuity, but also there would be a restoration of the experience of unity, another of the three major criteria for personal identity.

[22] Eagle ends his paper urging that his "main position . . . is that while most of us may be assured of numerical unity (that is, one self per body), as far the issue of integration is concerned, unity of self and of personality is a developmental integrative achievement in which we are not all equally successful".

[23] These two senses of the unity of the self correspond, respectively, to Eagle's "unity of the self" and "unity of personality", both from the point of view of the degree of integration achieved by a person.

REFERENCES

Ackerman, R. J.: 1972, *Belief and Knowledge*, Doubleday and Company, Garden City, New York.

Arlow, J. A. and Brenner, C.: 1964, *Psychoanalytic Concepts and the Structural Theory*, International Universities Press, New York.

Brody, B. A.: 1980, *Identity and Essence*, Princeton University Press, Princeton, New Jersey.

Danto, A.: 1963, 'What we can do', *Journal of Philosophy* 60, 435–445.

Danto, A.: 1965, 'Basic actions', *American Philosophical Quarterly* 2, 141–148.

Davidson, D.: 1980, *Actions and Events*, Clarendon Press, Oxford.

Dennett, D. C.: 1969, *Content and Consciousness*, Humanities Press, New York, pp. 90–96.

Dennett, D. C.: 1976, 'Conditions of personhood', in Rorty, A. (ed.), *The Identities of Persons*, University of California Press, Berkeley, pp. 175–196.

de Sousa, R.: 1976, 'Rational homunculi', in Rorty, A. (ed.), *The Identities of Persons*, University of California Press, Berkeley, pp. 217–238.

Eagle, Morris: 1981, Anatomy of the self in psychoanalytic theory', in Ruse, M. (ed.), *Nature Animated*, Vol. II. D. Reidel, Dordrecht, Holland, pp. 000–000.

Fingarette, H.: 1969, *Self Deception*, Humanities Press, New York.

Frankfurt, H.: 1971, 'Freedom of the will and the concept of a person', *Journal of Philosophy* 68, 5–20.

Freud, S : 1905, *The Psychopathology of Everyday Life*, MacMillan, New York.

Freud, S.: 1963, *Dora – An Analysis of a Case of Hysteria*, Collier Books, New York.

Freud, S.: 1963, *Introductory Lectures on Psycho-Analysis, Standard Edition*, 15 and 16, Hogarth Press, London.

Freud, S.: 1971, *The Ego and the Id, Standard Edition*, 19, Hogarth Press, London, pp. 12–66.

Goldman, A. I.: 1970, *A Theory of Human Action*, Prentice-Hall, Englewood Cliffs, New Jersey.

Kohut, H.: 1971, *The Analysis of Self*, International Universities Press, New York.

Kohut, H.: 1977, *The Restoration of the Self*, International Universities Press, New York.

Laing, R. D.: 1959, *The Divided Self*, Tavistock Publications, London.

Locke, J.: 1947, *Two Treatises on Government*, Hafner Publishing Company, New York, pp. 133–146.

Ludwig, A. M., Brandsma, J. M., Wilbur, C. B., Bendfeldt, F., and Jameson, D. H.: 1972, 'The objective study of a multiple personality', *Archives of General Psychiatry* **26**, 298–310.

Moore, M. S.: 1975, 'Some myths about mental illness' *Inquiry* **18**, 233–265. (Also in: *Archives of General Psychiatry* **32**, 1483–1497.)

Moore, M. S.: 1980a, 'Responsibility and the Unconscious', *Southern California Law Review* **53**, 1563–1675.

Moore, M. S.: 1980b, 'The nature of psychoanalytic explanation', Psychoanalysis and Contemporary Thought 3. (Also in: L. Lauden (ed.), *Mind and Medicine: Problems of Explanation and Evaluation in Psychiatry and the Bio-Medical Sciences*, University of California Press, Berkeley).

Moore, M. S.: 1980c, 'Legal conceptions of mental illness', in Brody, B. A., and Englehardt, T. (eds.), *Mental Illness: Law and Public Policy*, D. Reidel, Dordrecht, Holland, pp. 25–69.

Mullane, H.: 1971, 'Psychoanalytic explanation and rationality', *Journal of Philosophy* **68**, 413–426.

Nagel, E.: 1959, 'Methodological issues in psychoanalytic theory', in Hook, S. (ed.), *Psychoanalysis, Scientific Method, and Philosophy*, New York University Press, New York, pp. 38–56.

Pears, D. F.: 1975, 'The appropriate causation of intentional basic actions', *Critica* **7**, 39–69.

Prince, M.: 1930, *Dissociation of a Personality*, Longmans Green and Co., London.

Ryle, G.: 1949, *The Concept of Mind*, Hutchinson and Company, London.

Sachs, D.: 1973, 'On Freud's doctrine of emotions', in Wollheim, R. (ed.), *Freud: A Collection of Critical Essays*, Anchor Books, Doubleday, Garden City, New York, pp. 132–146.

Schafer, R.: 1976, *A New Language for Psychoanalysis*, Yale University Press, New Haven.

Schelling, T.: 1980, 'The intimate contest for self-command', *The Public Interest* **0**, 94–118.

Taylor, W. S. and Martin, M. F.: 1944, 'Multiple personality', *Journal of Abnormal Social Psychology* **39**, 281–300.

Thalberg, I., 'Freud's anatomies of the self', in Wollheim, R. (ed.), *Freud, A Collection of Critical Essays*, Anchor Books, Doubleday, Garden City, New York, pp. 147–171.

Watt, A. J.: 1972, 'The intelligibility of wants', *Mind* **81**, pp. 553–561.

Wiggins, D.: 1980, *Sameness and Substance*, Harvard University Press, Cambridge, pp. 149–189.

Williams, B.. 1973, *Problems of the Self*, Cambridge University Press, Cambridge, pp. 15–18.

Wittgenstein, L.: 1958, *Philosophical Investigations*, Blackwell, London, p. 161.

DAVID GRUENDER

PSYCHOANALYSIS, PERSONAL IDENTITY, AND SCIENTIFIC METHOD

INTRODUCTION

Professor Eagle has written a wide-ranging and provocative paper in which he attempts to come to grips with a number of philosophically basic criticisms of any personality theory that contemplates fundamental personal disorganization or unconscious or subconscious actions. Without allying himself with any particular school of thought in psychiatry or psychology, Eagle tries to show how these criticisms fail to deal with the problems of both normal and abnormal behavior by exhibiting a representative range of such problems for our examination and judgement. I think Eagle is fundamentally right, and I will offer arguments to support this position on methodological grounds, while, at the same time urging that existing personality theories are both too simple and not well-enough integrated with theories in related areas. The first will occupy the greater part of my remarks.

* * *

A very large portion of Eagle's paper is devoted to showing us something of the details of human experience that he thinks must be taken to count as reasons in favor of personality theories which recognize unconscious actions and disorganized personalities as common parts of the human condition. Indeed, Eagle concludes (with Ellenberger) that unity of personality is a human goal that is highly desirable but which, alas, may often not be achieved. The two issues are, of course, closely intertwined in that, granted the appropriate personality theory, one sign of personality disorganization is the unconscious action a person may take in conflict with a goal of his conscious action. However, on Eagle's view, unconscious actions are pervasive in human life and may be entirely benign.

It seems to me that the weight of Eagle's argument is decisive, but it is important to see that it turns on recognizing that the function of a personality theory is *explanatory*, in the scientific sense of that term. Hence the probative value of exhibiting the range of phenomena that occupy the bulk of Eagle's paper rests on the ability of the theory to account for those phenomena. And

Michael Ruse (ed.), Nature Animated, 203–222.

it is only by virtue of the fact that the theory does account for them that the individual phenomena can be said to be *reasons* that favor the theory or theories involved. In fact there are a variety of theories, but as they involve both the possibility of unconscious actions and mechanisms whereby the integration of a personality may or may not come about, we may conveniently treat them as a single family with respect to those philosophical arguments critical of such features. I make this point because those who have offered the philosophical criticisms see their contribution as conceptual, logical, or "analytic" (in the sense that this is something philosophers, not psychiatrists, do), while the reply is empirical. Yes, Eagle tells us, we cannot take psychoanalytic theories as conceptually perfect, but here are some facts they account for, and these cannot be ignored.

I think the philosophical critic believes he can enjoy the luxury of ignoring them for a variety of reasons, which I will treat in turn.

To start with one kind of argument that is stark and simple, consider the philosopher who says that there is no such thing as subconscious or unconscious thought because to suppose otherwise would be a contradiction in terms. Thinking is conscious activity; to be 'thinking' means to be aware that and of what one is thinking. Depending somewhat on the orientation of the philosopher, these are offered as 'conceptual truths', 'analytic truths', the consequences of unarguable definitions, common sense, or the result of reflection on ordinary language. The literature in this *genre* is so extensive that further amplification of the view is probably redundant for most readers. It is enough to note that it grew out of the work that Wittgenstein began to prepare for publication as his *Philosophical Investigations.* We need to recognize, too, that this same work might similarly be seen to excuse the philosopher from the need to explain or account for anything, on the ground that while science may call for explanations, the tasks of philosophy are fully discharged by making concepts clear. It may or may not be only a coincidence, but all the examples of clarification in Wittgenstein and in those who carry on in his tradition are of concepts that are taken from personal or social contexts.

Coincidence or not, however, it may tacitly make us more willing to accept the dichotomy between those situations in which our task is scientific and explanatory, on the one hand, and those in which it is philosophical and calls not for explanation but only a better understanding of our own concepts. Since, for most people, even in our own day, it is the physical sciences that are truly scientific, nothing seems to be lost by this move. It would merely remove the possibility of the social sciences, including psychology. We are all willing to grant astronomers and biologists and chemists and physicists, and such like, an exclusive license to explore and explain things in their own fields, but most of us think we know as much about people as anyone else and have little inclination to take the social sciences seriously, much less as a part of the enterprise of scientific inquiry. This is convenient, too, for those who consciously or unconsciously (if I may be forgiven for using these words), take the attitude that humankind, human nature, and human social life are sacred, special, or somehow beyond the natural order. Although this view may have religious orgins, it is by no means confined to the religious. I know a talented and irreverent physicist who enjoys warning his colleagues against transgressing what he reports to be the Eleventh Commandment: "Thou shalt not commit a social science".

The situation is not helped by the failure of the social sciences so far to have uncovered important elements of knowledge dramatically at variance with ordinary ways of thinking about ourselves. For a while it looked as though Freud's work had done just that, and that he was, perhaps, the Copernicus (if not the Newton) of the sciences of man. But with the growth of his theories confined to what is better described as a cult than a science, and with vigorous criticism of its basic approach by behaviorist psychologists who waved the banners of science so enthusiastically as they talked that some thought they marched in its army, respect for Freud is greatly diminished now. It largely remains, aside from members of the cult itself, with literary types who, thanks to what Snow has described as the phenomenon of our two cultures, are not troubled by doubts over its soundness as science. There are other factors as well. I think the most important is that the bulk of mental illness remains with us in spite of the application of his methods (and, to be fair, those of others), so that we cannot think of him as a Pasteur or a Salk either.

But whatever history may one day judge about the overall success of the sciences of man so far, it is understandable why so many would be confident that their knowledge of the race was as good as anything else available. Couple this with the religious view already mentioned, and it becomes painless to make conceptual moves that leave no room for the scientific investigation of *Homo sapiens*. The view that what problems exist in the area require no explanation; for nothing is hidden, only description is required and a perspicuous analysis of our common concepts and how we learn them – owed to Wittgenstein – is one of these. It has been applied and extended in this area by Ryle, Winch, Hamlyn, Malcolm, Donnellan, Shoemaker, Davidson, and many others. The issues raised by this move are complex and numerous, and it is not my purpose to pursue their exploration here. I want only to draw attention to the fact that, by ruling out in advance the possibilith of the explanation of human phenomena in any terms whatever, the possibility of the sciences of mankind is equally denied. So much for the grand hopes of Comte that the next move forward for humanity would be through the application of the scientific method to a better understanding of ourselves. While this consequence is of sweeping importance, it has not been overtly recognized, although some remarkable claims have been made under its influence. Consider, for example, Winch's portrayal of the social sciences as really a proper branch of philosophy. Or Hamlyn's assertion that psychophysics is 'conceptually impossible'. Nor were such views held merely as a quaint eccentricity without a cash value. My late colleague William Rushton, whose work was in human vision, once recounted his horror at the announcement of the same "conceptual truth" by Gilbert Ryle, with whom he then sat as a member of the University Grants Committee. The latter then proceeded to propose that such work no longer be supported. Fortunately cooler, and, Ryle must have thought, more muddled heads prevailed, and psychophysics survived in the United Kingdom. For years thereafter Rushton avoided contact with philosophers.

There is no question that the search for greater perspicacity and clarity regarding our current concepts about ourselves has been and will continue to be of long term benefit to all of us. Nor is this the place to evaluate Wittgenstein's programs for solving problems of philosophy. As a methodological matter, however, while clarity is always desirable, to limit the object of the clarification process to current and everyday concepts would bring scientific inquiry and the growth of knowledge to a halt. If we have learned anything at all from the history of science it is that new knowledge is won by changing our concepts of a field, and that this knowledge, as it is grasped, stimulates further such changes. In any field investigators must begin with the common understandings of society: be they that the world is flat, epilepsy is a disease whose victims are divinely inspired, or that a person is a unified and indivisible whole. But it is only a beginning. We tinker with the old concepts as we try to take account of what we observe. We look for patterns in the events and try to account for these with altered concepts, testing them against new observations. And as we ponder the shifting appearances we are sometimes able to invent new concepts to apply and test; sometimes with and sometimes without success. After a period of such scientific activity, the concepts will be different from those used at the outset, and our knowledge more extensive and reliable. It is this process we are told not to apply to ourselves and our lives, and it is important to see that I do not wish to be understood as asserting that this is what Wittgenstein meant. He left his writings in such a state that they are open to a variety of interpretations, and I am not among those who are positive about his intentions. But we do have some of his words, and I think the conclusion is a consequence of taking those words seriously.

That conclusion is reinforced by looking at another aspect of those words. Not only is his reader counselled against using other than the current everyday concepts, he is told that nothing is hidden and that problems in this area require no explanation. To the extent that pronouncement is taken to apply to ideas, concepts, problems, and puzzles about people and the complex social life they generate, to just that extent is the scientific explanation of human life ruled out. For our purposes it is not necessary to consider what reasons might have been offered in support of this view, or its role as a philosophical strategy. It is enough to recognize that this view would, as a matter of principle, rule out the use of science as a tool in understanding ourselves, and that whatever other benefits it may be thought to confer, it is fatally flawed in this regard. In fact, Wittgenstein offers no reasons for his view, and would object to them being offered on its behalf. But what is important is that no method exists for determining that some field is unfit for scientific inquiry. We can only apply that inquiry and see what skill and luck and determination can bring us over time. And that is what we must do with the problems of understanding ourselves.

I suspect that Wittgenstein, Ryle, and others were led in this direction by reflection on earlier conclusions that there is something radically private about each human being's individual experience. Add to this the recognition that language is ineradicably a social product, and constructed only to enable us to deal with one another in a shared world, and it might well seem that, as Wittgenstein once put it, as far as we can *say*, our experiences, like the beetle in a box we cannot open, make no difference and might just as well be nothing. All of this has been discussed extensively during the past thirty years under what has been called the "private language argument". It is an argument that, for our purposes, we need not enter, for its outcome is beside the point at issue here, although that may not be easy to recognize. That point is whether or not is possible to refer to, and to form and to test hypotheses about an individual's experience and thinking, conscious or unconscious. Why is this a different problem?

The answer, in the briefest possible compass, is that what Wittgenstein seemed to have wanted originally was a way to get at the ineluctable particulars of the experience of an individual. It is just these elements which existing social language cannot characterize without loss. Likewise, an individual may be thought to have such privileged information in that regard that no knowledge anyone else might possess could possibly have a bearing on it. So no one else would be in a position to test any claims an individual might frame even were he to have a language in which they could be expressed. And, finally, even if following Russell's suggestion, that individual might devise a thoroughly 'private' language for his own use, he could not even verify any of his own claims, for nothing would count as a test. Peirce would explain that nothing would count here because what would be expressed, if that were possible, would be entirely particular to that occasion, so that nothing that happened on another occasion could have a bearing on it, and the occasion itself, once experienced, ceases to exist, although it may be recovered *in part* through memory. Wittgenstein is satisfied to ask the questions and then let us follow him to the conclusion that there is no answer by our inability to provide one. If we are persuaded by this, reference to mental events, conscious or unconscious, would be impossible.

But our scientific interest in the experiences and thoughts of another person and in human behavior generally is not in the ineluctably particular facts of those experiences, which, as Dewey put it, can be enjoyed or suffered in their fullness but described only in abstract terms. Our scientific interest lies, rather, in the abstract way these experiences can be described, the repeating patterns they seem to display through the lives of individuals or groups, and the connections these have to our functioning as individuals in our shared world. A psychiatrist, psychologist, or sociologist, whether trying to understand our cognitive, affective, interpersonal, or other aspects of our life, are interested in what can be shared, what can be tested, what can be found to fit patterns in space and time, and what can be traced to other causes, be they mental or physical, genetic or environmental, physiological or psychological or social, or perhaps even logical. What is as good as nothing, like the beetle in the box, is not the object of this pursuit. Mental events, in the scientific context are, however successfully, dealt with in theories that tie them to things we can observe, are subject to test, and, being discussed in the languages we share, are designed to account for and explain things that puzzle us, and through their use enable us to understand and deal with our shared world. When Freud offers us a theory that accounts for certain kinds of forgetting, or Rushton for quantum phenomena in human color vision, both deal with classes of mental events they connect with our shared world, not those so radical in their particularity and privacy that they are unspeakable. Perhaps they are not very good theories, but their inadequacies lie not in their dealing with mental events, conscious or unconscious, but in how well they explain and account for the phenomenon in question, how well they interface with other theories in their own and related fields, and how well they stand up to repeated observation and testing. It is because they have this character that the arguments of Wittgenstein (and those derived from his work) that would appear to remove mental events from what is discussable are beside the point at issue here. The point is epistemological and methodological, not ontological. Had anyone attempted to identify knowledge of the physical world with that of the unique and the particular – for which there is equal reason – the result would have been the impossibility of the physical, rather than the social sciences. And it would have been equally unsound. Here I use 'mental' and 'physical' in the normal phenomenological sense, but (*pace* Quine) with no ontological commitments. That is to say, there are phenomena we may think of as 'physical' and others as 'mental', while remaining open-minded as to what the ontological status of these two categories may be.

Yet it is because of the misunderstanding of this epistemological and methodological point that one may think that mental events, conscious or unconscious, are nothing to be explained, for nothing is hidden, and that the only cognitive activity available to satisfy our curiosity on this head is to clarify the implication of our common language in this area. This rules out unconscious thinking as an unclarifiable monster, and generates the supposed impossibility of the scientific study of humanity. It would appear, however, that this study is merely difficult, not impossible.

* * *

Although he does not cite them, there is another source from which criticisms of positions like Eagle's have come, and fairness requires us to consider it. I mean the traditions of logical positivism or logical empiricism, and, to the extent they shared a common methodological base, those of operationism and behaviorism. There is an enormous literature on this subject, which we need not here review fully. For our purposes it is sufficient to look at the claim that concepts that refer to mental events are not operational or behavioral or verifiable, and hence are not meaningful. Whether in the hands of Bridgman, Skinner, or Carnap, this view does not rule out the scientific study of mankind, but prescribes its form. The resulting form would then make it impossible for us to offer theories about conscious or unconscious mental activities. A vexing problem with these approaches, as had been noted by many, is that no one has succeeded in clearly and adequately articulating what is to count as "operational", "behavioral", or "verifiable", although major efforts in this direction have been made by many for more than fifty years. Another is that there is nothing transparently epistemologically virtuous about "operations" or "behavior". They are taken to be observable. If we pass over the many problems that would have to be faced in evaluating this claim, we would come face to face with the epistemological claim that one could rightly hazard theories only about what was observable. Fortunately Lavoisier did not have to wrestle with such views. As Skinner would have it, no theories are necessary at all, thereby stretching the meaning of 'theory' and 'observation' to the breaking point and beyond.

The concept of verifiability does require more serious reflection, for it is really an attempt to generalize and make rigorous the widely accepted recognition that science, as a method, enables its practitioners to advance in their knowledge of nature by striving continually to test its theories against experience. When, in some case such a test has been carried out successfully, whether with positive or negative results, we can say after the fact that the theory has been tested and thereby at least partially verified or disverified. The original vision was to invent a method whereby the syntactic and semantic means for doing this could be specified, so that, once they were understood in general terms, one could then turn about and specify *before the fact* whether some theory was verifiable in principle or not. One would need merely to apply the syntactic and semantic tests. The classic formulation is that of Carnap. The syntax was that provided by propositional logic with the uncomplicated semantics that a simple or 'atomic' proposition had the value of truth or falsity. Since all other propositions more complex than these, the 'molecular' propositions, were formed only from the former and logical operators, they were truth-functionally dependent on the atomic propositions. On this scheme it looked as though one could look at any theoretical claim in science and, through a process of logical analysis, determine the atomic sentences whose truth-values would settle the issue. While these were first thought of as being primitive sensory experiences of the kind Wittgenstein later took to be radically private, that idea was dropped in favor of a physical interpretation when the problems of privacy became evident. Although Carnap himself maintained a principle of tolerance regarding this choice of language, physical interpretation remained dominant. And, over the years, serious syntactic, logical, and semantic problems have given rise to various modifications of the scheme. Its relevance to our problem is in the physical interpretation of the atomic propositions, or 'protocol sentences', as they were later called. For on this interpretation we have a philosophical theory (or metatheory) about science which confines the testing of a scientific theory to the observation of physical phenomena, together with a set of linguistic standards that, essentially, limits the scientific theory to stating logical operations to be performed on these. As a result of adopting such a metatheory, we would prevent ourselves from having any theories about mental acts or events, conscious or unconscious, much less about the relative integration of persons. As in the case of Wittgenstein's view, no problem of the explanation of any mental act or pattern of such acts can arise, for what falls outside the metatheory is designated as unfit for scientific inquiry because it is 'meaningless'. Had he adopted this approach, Freud could have learned that he had no problems requiring explanation or inquiry.

But what reasons are there for adopting such a metatheory? This is not the place for a full evaluation of the reasons for and against doing so, but a few things can be said briefly. First, historically, none are offered. Those who worked on the metatheory have been content to develop each version of it to surmount the logical difficulties found in the earlier ones, but from first to last have been content in the thought that they were merely explicating 'meaning' and 'science'. This is deceptive. A theory of meaning as much stands in need of reasons for our adopting it as any other. One can develop a complex and abstract system and offer it as a theory of meaning, but in the end a scientific investigator needs to know whether it is a satisfactory theory, whether it accounts for the phenomena in the field, whether it fits other theories where it might be expected to, and so forth. That task cannot be avoided by remarking that one is merely explicating a well-known concept, for there are explications and explications. Even the explication of a canonical text by a religious authority does not escape challenge in our day. Unfortunately, the same has not been true in philosophy. The task of justifying the metatheory has been left undone, with the result that those who adopted the metatheory have acted, in effect, as though they had a proprietary right to the concepts of 'meaning' and 'science'. But while this looks like dogmatism from the outside, the inner conviction that one's abstract scheme was just a clarification or explication enabled one to avoid this recognition.

Are there good reasons for the adoption of any such schemes? I think so, but they seem to me to be outweighed by the exclusion of large areas from scientific inquiry, again particularly in the social sciences, and by the syntactic limits they would impose on the methods of science. Both are arbitrary. The methods of science are as little subject to precise delimitation as is its substance. And, in this regard, another serious problem arises even if we ignore the special difficulties of each version of the scheme. The problem is that the scheme began with the vision that the essence of science lay in the empirical testing of its hypotheses. *After* an inquiry one might know whether one's hypothesis has been *verified* or not. But the point of logical empiricism is to determine *in advance* whether some hypothesis is *verifiable* 'in principle'. Not remarkably, Carnap traces this idea back to Wittgenstein. The general plan would make this task conceptually easy by analyzing the very *meaning* of the hypothesis into its components, of which the empirical elements, stated in the 'atomic sentences' or 'protocol sentences', would consist in the experiences one would have or physical states of affairs one would observe were the hypothesis true. To determine whether some hypothesis is verifiable in advance of our attempting any test of it, we would need merely to put it in proper form and look for those atomic sentences. If we find any and they fit the protocols, then our hypothesis is meaningful and verifiable and inquiry may proceed in normal scientific fashion. For our purposes we may ignore the problem that, if you find any at all, they will be in endless number. Our problem, rather, is that unless you could say, quite particularly, what those protocol sentences describing possible experiences or observable states of affairs are, your hypothesis would come up as meaningless. And yet, except in the most routine investigations, one s hypotheses do not appear to have such a content and even that appearance is illusory.

If we look at the specification of these possible protocol sentences, they will in fact be found to cover a wide range of degrees of abstractness. At the least abstract, where there is no degree at all of generalization, we have experiences or situations that are too particular to be characterized or expressed in language. They can be enjoyed or suffered, but there is nothing that can be said, and *a fortiori*, nothing that can be captured in a protocol sentence. Approaching the opposite extreme, we can have a hypothesis that we think would make some difference in our interaction with the world, but are not able to characterize it. In between lie most of the cases, in which we can specify or recognize the possible observational outcomes to some degree. It is in these cases that judgment, experience, and a wide variety of theoretical considerations enter that cannot be specified in advance, least of all embodied in a set of logical and syntactic rules. This is masked from our sight when, to check the application of the metatheory, we try it on a low-level hypothesis that has been accepted for centuries: say, that copper expands when heated. We already know it to be true and are confident that if we have an analysis of its meaning it will at least contain the elements that the copper after heating will be measurably longer than before. The situation is radically different when we are working at the forefront of knowledge. How much longer? Measured how? Heated how? Heated how much? How many times? How many samples of copper? It is only because all these – and other related – matters are settled and familiar that we can imagine an outcome has been specified. With the general theory of relativity, of which there are now a good many varieties, we appear to be perhaps a decade away from having developed the necessary equipment ot begin to test the various consequences of each variety, and far from being able to specify precisely what we expect to observe in each case. Indeed Einstein, the author of one of these, died knowing even less about these hoped-for observations than we can know today. Or contrast the situation with the hypothesis that some mental illness is the result of the defective production or metabolism of endorphins in the brain. We do not yet know how these substances are produced nor how they function. We can imagine experiments which supply analogous substances – morphine and its relatives – to sufferers and see whether there is any relief from the symptoms, but that is pretty rough and compatible with the falsity of the hypothesis. At the same time we are in no position to specify what we will find or observe as we do learn more about the structure, function, and physiological history of the endorphins. Only the future progress of investigation can yield such knowledge.

And that is the general difficulty. Verifiability is presented as a meta-theoretical syntactic property that can be determined in advance, but when one looks at the subject matter of which this is proposed as the metatheory and at the consequences of the metatheory itself, it turns out not to be the case. There are, of course, many other factors that deserve consideration in this matter: the difficulty of determining the implied consequences of a hypothesis, for example, especially when they are endless. But enough has been said, I think, to show that verifiability, like truth, is not among the properties of a hypothesis that can be determined in advance of a scientific explanation. Indeed, the only way to decide whether a theory or a hypothesis is verifiable is by trying to verify it. Valuable as logic and syntax are in science, they cannot serve as a substitute for empirical investigation.

I conclude that the two chief philosophical traditions that would appear to raise methodological bars to theories about mental events and acts, conscious or unconscious, or theories about the structure of personalities, are, whatever their merits, mistaken in this regard. There is, therefore, no reason for us to shrink from trying out or inventing such theories, nor to avoid testing, extending, or altering them in accordance with what we find and what our luck, skill, determination, and imagination may lead us to.

* * *

In such a spirit, and with an eye to methodological problems, let us now look at some of the detailed issues Eagle describes with the hope that such theories may be seen to fill a genuine scientific and explanatory role. Mental events or acts – thoughts in the generic sense – are a good place to begin. That each of us is privy to much activity, namely our own, and, can know something of its connection with our speech and writing and other expressive behavior, as well as our actions, and its relation to things that happen to us and that we observe – is unremarkable. That most other people seem similarly equipped would account in a general way for what we can observe of this behavior and our own interactions with them, and is equally unremarkable. Nothing deep need be implied in either case, and we may revise our views on the basis of more evidence. This is not science, but it is not nonsense either. Now, what of the hypothesis that there is mental activity we are not privy to: subconscious or unconscious thinking, cognition, perception, emotional response, planning, calculating, even bodily acting?

The first objection to be recorded is that this would be a contradiction in terms: 'thinking' means 'what one is conscious of'. To support this contention the witness of common sense or ordinary language or classical authorities may be summoned. Or sometimes the point is merely put archly, with the implication that any reader of discernment and intelligence would agree. Alas, these are all forms of the *argumentum ad verecundiam*, and no less shabby for having been stated by eminent philosophers. It might help here to remind ourselves that one of the arguments given Galileo against his view that the earth revolved around the sun was the claim that 'earth' meant 'stable and immovable body at the center of the universe'. Of course there would be a contradiction in terms if the definition of "common sense" were accepted. But in scientific inquiry definitions make sense only within the context of some theory, and the adequacy of any theory is always open to question. Definitions may be altered to serve any functional theoretical purpose. In the end it is the theory that must stand up to tests. Definitions do not stand alone. Nor could a definition from one theory stand as a bar to a different theory. However satisfactory the definitions of ordinary language may be to our communicating with one another on an everyday level, that is all we can take them to be. New knowledge requires new theories, and new theories will define their terms differently. The science of humanity is here no different from that of the planets: the deliverances of ordinary language and common sense have the same relevance in both.

There is, therefore, no defensible conceptual or syntactic reason for us to avoid considering theories that suppose there to be unconscious or subconscious mental activity. What are the advantages to such theories? Eagle lays out some intriguing ones: the concept would account for a wide variety of puzzling human phenomena, and could be ramified and developed to deal with details, as well as broad categories. Sometimes people do things but do not understand why. That is to say, they do not understand their own behavior, its motives, sometimes even its means. In fact, they may not be aware that it was their behavior until they later find incontrovertible evidence of this fact. "I was not aware of it – I must have been out of my mind!" Sometimes these actions are minor; sometimes they are fairly complex. But they are all of the kind that normally require cognition, deliberation, forethought, planning, and sometimes motor skills with continual monitoring. They may occur only rarely, or often in certain circumstances, or they may, as in multiple personalities, become a compelling, disturbing, and pervasive feature of life. A person may get the feeling he is losing control or being victimized by such episodes, and seek help. Approaches which use concepts of subconscious thinking try to provide that help by devising detailed theories of why the victim's mental processes have become opaque to him, and thereby seek means to their lucidity and control. Without commenting on the details of such theories, Eagle only wants us to see that they are at least the beginning of an explanation of these phenomena, which cannot themselves be responsibly brushed aside. I agree. The concept merely requires the separate recognition of mental activity on the one hand from our awareness or consciousness of it on the other. The mental activity plays a cognitive, affective, and integrative role in either case, on this view, with the difference between conscious and unconscious thinking being that, in the former case, the thinking is attended to by the person doing it.

The explanatory advantages of this move are wide but not deep. They enable us to understand many phenomena, from Eagle's involuntary leaning into a stalled escalator to some of the complexities of cases of apparent multiple personality, as varieties of thinking done beyond the focus of direct attention. When one looks at the data coming from studies of the specialized abilities of the cerebral hemispheres, including patients in whom the capability of communication between hemispheres has been impaired through accident or surgery, it is striking how much of what we take to be thinking goes on without our direct awareness. There must be mechanisms for focusing our attention on what, at any moment, is most important. And it does not seem unlikely, therefore, that such mechanisms could similarly function, in pathological cases, to direct our attention elsewhere; nor, given special needs and circumstances, to areas in which it is normally never needed. Something of the latter seems to be happening when we use biological feedback training mechanisms to solve special problems with bodily processes normally handled by the involuntary system.

Indeed, our biological history and complexity would suggest that most of our cognitive, affective, and integrative thinking must function much of the time without our direct awareness of it. Having arrived in this place, I am sure my feet brought me here, but what they were doing on the way was at or beyond the outer limits of my attention and conscious thought. I suppose I would have noticed them quickly enough, although probably not in time, had I stumbled. I will never forget my Grandfather's wry admonition to watch *both* my feet when, as a boy, I would come home with cuts or bruises on one of them. But I could never take his advice, nor can anyone else; which is why he gave it, for he was, after all, a philosopher of a sort himself. I can remember one of my earlier piano teachers being terribly exercized about how I held my elbows. I was, of course, not aware, except in the most general sense, that I had elbows. After a few days awkward attention to their new position in relation to the keys, I got used to it, and they promptly passed out of my ken. This newly regained blissful ignorance about my elbows gave me time to pay attention to my fingers. But then I noticed that, having mastered the finger techniques of that particular piece, I no longer paid attention to what my fingers were doing either. They became like my elbows, as far as my attention was concerned, and flew over the keyboard almost detached from me. What I attended to was the shading of dynamics and tempo and the expressiveness of the music.

Then there are other tasks that one can do so well that, after they have been started, can be pretty much ignored. I can remember going out to split some firewood on a Sunday afternoon while thinking intensely about another problem. Suddenly I startled upon seeing one of my children in the doorway of the barn speaking to me. Up to that time I had been aware only of my thoughts about my problem. The child had been standing there for some time watching me split the logs, and then somewhat hurt at not having been greeted, had spoken to me. I took her in my arms to comfort her, put my problem aside in my mind, and it was then I noticed with surprise the huge stack of logs I had split. It seems I had spent the whole afternoon at the task, not noticing the fading light or the calls from the house. My impression was that I had barely begun; in fact the only thing that had barely begun was the search for a solution to the problem I had entered the barn with.

On another occasion I can remember driving up to the front door of my home, turning off the ignition, and then asking myself where I had been and how I had gotten there. I had, again, been thinking about a problem and had focussed my attention on that. So I had to sit there in the front seat and try to remember where I had been and what I had done, so as not to appear an utter fool when I walked in the house. And, in the main, I did succeed in recollecting the significant events. But I could not, for the life of me, remember the route I had followed to get home, much less the events of the trip. Yet, assuredly, I had decided which streets to travel on, had waited at traffic lights, maneuvered through the press of vehicles, and had arrived safely at home – all tasks requiring perceptual judgement, motor skills, rapid responses, and deliberation. I presume they were exercised, but not with my conscious attention, for that had been fully engaged with my problem.

I do not place great significance on these isolated and idiosyncratic accounts. Whatever weight they bear comes from the extent to which others had similar experiences, as well as their appearances in literature generally. They happen to be very widespread.

It is as though the complexity of our abilities as human beings is such that the gift of conscious attention is reserved for only a fraction of our activities: those which seem to need it most then. The bulk of what we do, including much of our thinking, can be left to more routine processes. We could not consciously process all the information necessary to walk two steps in time to take those steps. But once having learned to walk, we could focus our attention on something else. Such an ability would have great survival value in creatures as general as ourselves, and I suspect, therefore, that in the normal case much of our thinking is unconscious or subconscious, and it is only our understandable attachment to and identification with the conscious fraction of ourselves that could occasionally lead us to ignore the rest.

But however wide the explanatory power of this concept may be, it is not very deep. For we have little, yet, in the way of more complex theories that attempt to account for the content of thinking, conscious or unconscious, or for the multitude of its connections with physiological, psychological, historical, social, and other factors that influence our lives. It is the development and testing of such theories that will, in the end, prove the value of this approach.

* * *

That brings us, naturally, to our concepts of ourselves as persons. There is a body of literature on this topic in philosophy, although it is modest in size compared to that in psychiatry and psychology. There is a much richer body of philosophical work exploring the problems of human agency and action and it too has a bearing on this concept, for we like to think of ourselves as the unique individuals we are in terms of our past actions and possible future ones. That may only be a reflection of our current culture, for human beings have thought of themselves in different ways throughout history and across societies. However, I shall concern myself here only with the issues around conceiving ourselves as unitary, single, and single-minded individuals.

What, for example, are we to do with a theory like Freud's which postulates a self as the result of the mutual interactions of such portions of a mind as the 'id', 'ego', and 'superego'? This gets us nowhere, Eagle finds Thalberg saying, for each of these imagined little elves in the mind is then itself a person, and we shall have to assign each of them a tripartite soul, and so on, with the effect that we shall never have explained anything: a modern use of the "Third Man Argument" from Aristotle. But this is too simple a reading of Freud, I think. He does not treat these as separate persons, but as separate forces within a person. And here I must applaud Eagle's suggestion that we all agree to adopt the rule that we speak of only one person to each human body. As a rule it may save us from type errors of just this kind. Theories postulating diverse elements struggling within a person may suffer from various disadvantages that further investigation may unfold, but their statement is not merely the error of taking them as examples of the type they are to explain. Freud, with such a theory, does no more than explain molecules in terms of atoms.

Indeed, such theories have an ancient and honorable history. Plato, too, suggested a tripartite soul, with a 'vegetative' element interested in self-nurture, a 'spirited' element interested in adventure and self-aggrandizement and a 'rational' element whose task was to harness and harmonize the other two in the interests of providing a good life for the person in a society of similarly constructed persons. That the function of the 'id', 'ego', and 'superego' are roughly parallel has been remarked upon before. My purpose is not to harness Freud's chariot to Plato's authority, nor even to point out that such ideas have been with us a long time and are part and parcel of the culture, not something new and strange. It is, rather, to emphasize that we have travelled only a little way with them in a very long time. There is, methodologically, nothing wrong with explaining persons on the basis of such theories. But until these elements are tied to physiological, social, genetic, psychological, developmental, cognitive, and other relevant variables, we will not have progressed much beyond the stage of myth. Freud made a start in this direction, but the general state of our knowledge in this area was not able to support much. We are in a better position now, and it is time to strike out in new directions. And to take such steps I suggest we will have to cross the conventional borders separating medicine, biology, psychology, and sociology.

However that develops, suppose it be granted that our psychological make-up is complex. We have all experienced conflicting impulses, and perhaps talk about a 'tripartite soul' is only a way of dramatizing that. Nevertheless, we resolve those conflicts before we act, and it is in that resolution that the real unity of our personality lies. That is why we are prepared to be judged as persons by our actions, and that is why we cannot conceive of ourselves as other than a consistent whole.

Something like this argument might be made by one who granted that the wholeness and essential unity of personality could not be established by definition or conceptual fiat. It might, perhaps, be called the 'legal argument', since it expresses ideas important in the Anglo-Saxon legal tradition, and which have played an important role in the philosophical literature since Locke. While I think Locke's discussion of this topic has been widely misunderstood in the last twenty years, the importance of this general move for our purposes is that it brings the issue to an empirical and factual level where Eagle's arguments apply. And their purport, to pass now over their details, is that while sometimes people resolve their conflicts and are prepared to stand behind their actions, sometimes they do not. The former may well be a praiseworthy goal, an idea for all to emulate, and a model toward which therapists may seek to move their conflict-ridden patients (or clients, depending on what one thinks of the medical model for mental illness), but it is simply not reflected in the actions of many all or some of the time. And it is these facts that one can begin to account for by theories that postulate mental structures with a dynamics of their own. What Locke, Charcot, Janet, and Freud do is open the possibility for a person who has done something inconsistent with that he *later* takes as his person to deny that the former act was his. Immaturity, lack of experience, personal instability, inconsistent behavior, or profound and unresolved personal or social conflicts may all play a role in this. And since a good part of our thinking or active mental life may escape our full conscious awareness, conflicts may not only remain unresolved, but different aspects or tendencies of a person may polarize around these. In this manner one may be tempted to think of an individual who displays a markedly different personality at different times and who is unaware at one time of the actions taken at another – as a case of 'multiple personality'. Such cases have been described since the earliest times. Until recently they were accounted for by postulating that the victim was possessed by an evil spirit. or, at any rate not by himself. Thus we may say, about our own temporary aberrations, that I was not myself, that I was beside myself, or that I was out of my mind. By the turn of the century witchcraft was no longer popular and possession by alien spirits was an explanation no longer socially available to the patients of the American psychiatrist Morton Prince. So these patients merely displayed multiple personalities, as have others, (sometimes to great popular interest) from time to time since. *The Three Faces of Eve* appeared even as a movie. These are facts about people we cannot afford to overlook. As Plato and Ptolemy remind us, our explanations can never afford to forsake the phenomena. When such phenomena are seriously taken account of, along with those of the development of children into adults, the darker side of mankind described and prescribed for by Sartre and Heidegger, and the large literature of what used to be called 'abnormal psychology' but which now seems to fit the mad world around us almost as well – all of these, it seems to me, only make more plausible Ellenberger's and Eagle's suggestion that unity of a personality is an ideal to be striven for with most of us. Certainly there are outstanding examples of human beings whose strength of character and achievements we admire. But if we learn of their internal struggles, that comes later, and we are more apt to think of the pinnacle they reached than the process by which they got there.

* * *

Throughout much of this essay I have tried to point out that the critics of dynamic psychiatry have overlooked the importance of explanation and why. Unfortunately, I think it is also true that the psychoanalytic tradition has also undervalued explanation, and its failure to move forward may in part result from this. My purpose here is only to illustrate this contention, not provide an exhaustive catalog, in the hope that a greater sensitivity to the methodological importance of explanation can serve to improve our practice.

One problem area with psychoanalysis is the narrowing of the field to those who have undergone analysis at the hands of a proper practitioner and are willing to spend their lives analyzing others. While its justification was to insure professionalism, Freud's own later behavior was as the jealous guardian of a religious sect, not the leader of an open community of scientists, and the procedure smacks more of the "laying on of hands" than of scientific training. Research is virtually eliminated as a professional goal, and ties to disciplines with related knowledge are not strong. Likewise, the obligation to account for observed facts rests but lightly on psychoanalytic shoulders. Most of the published material is in terms of descriptions of clinical cases, with but little interest in attempting to account for numbers of these with theoretical elaboration and then test those accounts against new data. Statistical and probabalistic considerations almost never appear. Yet any attempt to account for phenomena with that range of variability would require them.

Another difficulty in this field lies with the restrictions placed on the gathering of data. Psychoanalytic data are gathered and analyzed by the therapist in discussion with the patient. There is no interest in the independent checking of such data with records or other parties. Given Freud's sensitivity to the danger of the patient and therapist falling into unconscious deception of one another, this one-sidedness is especially regretable. There is also, in this arrangement, no means for checking the interpretation and judgement of the therapist. An explanation in science should stand the test of investigation by anyone qualified.

Finally, the very mechanisms of psychoanalytic theory hold the potentiality of explaining too much. If a patient resists an analyst's interpretation, the analyst is trained not to be surprised: the theory calls for the patient to resist; this sort of truth is not pleasant. But, by the structure of the therapeutic situation, how would an analyst discover that he was mistaken? By talking it over with *his* analyst, who is at an even greater distance from the facts? I do not recommend that the patient be taken at his word either. I merely wish to remind us that we can structure the therapeutic situation in such a way that no tests of the validity of the theories being used are possible, and in that situation we are all deluded should we take them as accounting for anything.

But all of these defects are reparable in principle. Psychiatry is willing to face unhappy facts about mankind, and has a few theoretical tools to use. Our experience with therapeutic drugs for mental illness is beginning to touch our growing understanding of brain function and metabolism. And work in cognitive psychology and genetics is reaching areas which overlap these. Add the hope for more sociological knowledge bearing on these problems, and I think we may see that the stage is being set for some new and solid achievements in our understanding of ourselves. Perhaps this attempt to exorcize some ghosts of the past may help.

Florida State University

PART V

W. F. BYNUM

THEMES IN BRITISH PSYCHIATRY, J. C. PRICHARD (1786–1848) TO HENRY MAUDSLEY (1835–1918)

1. INTRODUCTION

Viewed within the European context, nineteenth century British psychiatry was something of an intellectual backwater. Ackerknecht's *Short History of Psychiatry* mentions only five nineteenth century Britons, three psychiatrists, one neurologist, and one layman. Karl Jaspers' brilliant historical appendix to his *General Psychopathology* concentrates exclusively on French and German psychiatrists. And Professor Ellenberger's monumental *Discovery of the Unconscious* centres primarily on great Continental cities – Vienna, Paris, Zurich, Berlin – or on New York and the other American ports of call where not a few early pioneers of dynamic psychiatry settled. In the West, modern psychiatry, both the 'university psychiatry' of Kraepelin and Wernicke, and the dynamic psychiatry of Janet, Freud, and Jung, was largely created in Continental Europe. Even in the present century British psychiatry has benefited from imported talent, for arguably the two most distinguished practitioners of university psychiatry and psychoanalysis in post-war Britain have been Sir Aubrey Lewis and Anna Freud, born in Australia and Vienna respectively.[1]

This is not to denigrate the native British contribution to psychiatry, nor to suggest that the history of the subject in Britain is little worth studying except for its parochial interest. Certainly there was a flow of ideas and influence in nineteenth century psychiatry, and Britain exported to America and the Continent as well as receiving imports from those localities, as a number of translations, citations, and foreign visits attest. Nevertheless, I believe that we can understand the particular development of psychiatric theories and practices, and the psychiatric profession in nineteenth century Britain only by first grasping the ways in which that development was rooted in British medicine, and perhaps more importantly, in the religious, philanthropic, and cultural values of British society. It should be recalled that while Continental thinkers such as Comte and Weber were developing sociology, the British were busy perfecting the idea of social work; that while the Continental Intellectual has been identified with theoretical and systematic pursuits embraced under the rubric of *Wissenschaft*, the Intellectual in the

Michael Ruse (ed.), Nature Animated, 225–242.

land of Francis Bacon and John Locke has by and large been content with more limited, empirically grounded pursuits; that most nineteenth century Britons were suspicious of what they saw as narrow, abstract specialization, and that British doctors were proud of the pragmatic, utilitarian, and practical dimension of their profession.

These British characteristics can be dismissed as shallow and amateurish, or they can be defended as a genuinely positive, rich, empirical tradition which produced one of the greatest intellectual achievements of the nineteenth century: *The Origin of Species*. But whatever our attitude towards the comparative worth of what one historian has called 'the peculiarities of the English',[2] these peculiarities are surely related to the fact that the two most richly discussed themes in nineteenth century British psychiatry were *moral therapy* and the *non-restraint system*. Both were overwhelmingly practical issues, and while the former was independently though not uniquely British in its origin, moral therapy provided the conceptual underpinning for the development of the asylums with which so much Victorian psychiatry was associated. After a brief examination of the initial elaboration and ramifications of moral therapy and the non-restraint system, we shall look at the ways in which the values reflected in these themes continued to dominate psychiatry in Britain until the 1870s, when, within the asylum movement itself a new ethos began gradually to emerge.

2. THE SOCIAL MEANINGS OF MORAL THERAPY AND NON-RESTRAINT

Moral therapy has had no lack of historical attention; indeed, it is customary to date the birth of modern Western psychiatry from the efforts of Chiarugi in Italy, Pinel in France, and the Tukes in Britain.[3] The word 'moral' both in English and in its European cognates, meant more to these late eighteenth and early nineteenth century reformers than simply 'psychological', though generations of post-Freudian historians, attuned to the idea of the 'talking cure', have sometimes emphasized this aspect of moral therapy.[4] In its historical context, however, moral therapy was often contrasted to *medical* therapy and in this sense could include virtually everything except the administration of drugs, bloodletting, cupping, and other standard remedies which were employed for many disorders, and not simply psychiatric ones. In practice, it came to include education, work, interpersonal interactions and attempts at gradual re-socialization, and is the natural ancestor of contemporary behaviour therapy rather than psychoanalysis.

As initially developed by Pinel, from 1794 at the Bicêtre, and shortly afterwards at the Salpêtrière, and by the Tukes at the York Retreat (opened in 1796), moral therapy largely replaced medical therapy in those institutions, since on the basis of experience as well as for other reasons, both Pinel and the Tukes came to doubt the efficacy of medical remedies in the treatment of insanity. As Samuel Tuke wrote, 'the experience of the Retreat . . . will not add much to the honour or extent of medical science. I regret . . . to relate the pharmaceutical means which have failed, rather than to record those which have succeeded.'[5] More dramatically, though, the new moral therapy replaced not just medical remedies but the chains, whips, and other forms of physical restraint and coercion which had been common in the late eighteenth century. But as Michel Foucault has insisted, Pinel and the Tukes were as concerned as had been their predecessors to control their charges. The power structure in the institutions had not changed but the method had: in the new therapeutic environment, control was to be achieved by the altogether subtler means of moral therapy. As Foucault has put it, 'A purely psychological medicine was made possible only where madness was alienated in guilt.'[6] At the same time, the desired goal of therapy was to enable the patient to gain control of himself, for with this new therapeutic movement came an optimism about the curability of madness.[7] The patient's environment assumed such importance that, from the early nineteenth century, beginning particularly with Pinel, most writers on insanity devoted a great deal of space to the details of asylum design.[8]

The other significant aspect of moral therapy is the extent to which it coincided with new definitions of insanity, and in particular, the notion of *partial* insanity, elaborated by Pinel and generally accepted by French and British authors, though less so by some German psychiatrists such as Griesinger.[9] There was no logical connexion between the efficacy of moral therapy and the idea that insanity need not involve a total eclipse of the reasoning faculty, but the faculty psychologies of, first, the Scottish common sense philosophers, and, second, the phrenologists, reinforced the belief that insanity could be partial and that the lunatic could still be reached through his undamaged faculties. The idea of partial insanity thus increased the therapeutic expectations of the relatively optimistic early nineteenth century psychiatrists.

In British, these new notions of the nature and preferred treatment of insanity found physical embodiment in the Retreat, established by the Quaker philanthropist family named Tuke. The Retreat achieved national prominence in 1813 when Samuel Tuke, grandson of the founder, published his *Description of the Retreat.*[10] The book was widely and favourably reviewed and, in 1815, Samuel Tuke, his grandfather, and a number of other laymen active in the reform of facilities for the insane, presented testimony to a Parliamentary Committee inquiring into the conditions of madhouses in England and (in 1816) Scotland. The evidence, published by the Committee, seemed to establish three propositions: first, that moral therapy was associated with the best in the care of the insane and was both more humane and probably more efficacious than medical therapy; second, that doctors who had been in charge of various establishments for the insane had been in many instances guilty of neglecting their patients; third, that specialized asylums for the insane were desirable, particularly if these asylums were run along the lines established by the endeavours of the Tukes and their allies. At the time, most insane paupers in Britain were still confined in general workhouses or poor houses, even though an Act of Parliament passed in 1808 had given counties permission to erect, at public expense, specialized psychiatric asylums.[11]

These events were to shape the character of British psychiatric debates until mid-century, for doctors with a vested interest in the treatment of the insane felt threatened by the nature of the lay reforms achieved by the Tukes. Through a variety of activities, including a considerable literary output, public lectures, pressure groups, and, by the 1840s, a professional association and specialized journal, they worked to establish a disease concept of insanity located in the brain (rather than the mind, which was still frequently equated with the theological soul); to assert their own professional rights as the primary diagnosticians and therapists in cases of insanity; and to convince the ruling elite that public asylums, under the charge of a doctor, were worthwhile public investments. It is within this professional context that the achievements of Robert Gardiner Hill (1811–1878) at the Lincoln Asylum and John Conolly (1794–1866) at the Hanwell Asylum, near London, must be seen. Gardiner Hill began abolishing all mechanical restraints shortly after he became resident medical officer to the Asylum in 1835 and by 1837 he had effected their complete abolition. Conolly achieved his reforms beginning in 1838. But Gardiner Hill, though medically qualified, saw himself working within a humanist tradition, whereas Conolly saw non-restraint as the ultimate *medical* achievement within the asylum, a system of total environmental care aimed at restoring to the patient that loss of self control which was at the heart of his disease. As Conolly put it, 'the mere abolition of fetters and restraints constitutes only a part of what is properly called the non-restraint system. Accepted in its full and true sense, it is a complete system of management of insane patients, of which the operation begins the moment a patient is admitted over the threshold of an asylum.'[12] Not surprisingly, the still weak psychiatric profession applauded Conolly's efforts while turning a cold shoulder to Gardiner Hill, whose activities actually seemed to minimize their own claims to professional expertise.

Conolly left Hanwell after a few years in order to establish a lucrative private practice, though he continued to visit the institution in his capacity as consultant physician. More importantly, he continued to turn out a stream of books and articles which defended non-restraint. By the 1850s, when the asylum movement was in full swing as a result of the 1845 Act which required each county to provide one, British psychiatrists could look upon the combination of moral therapy and non-restraint as genuinely indigenous, humane, and therapeutically sound. They also saw it as peculiarly adapted to Britain, with its well developed tradition of individual liberty and toleration. As one psychiatrist wrote, commenting on the fact that Continental psychiatrists had not picked up non-restraint to any degree, it would 'be folly to expect that the merits of the non-restraint system should be recognized [in Germany] where even the sane portion of the community are drilled into order by soldiery and the police.'[13]

Thus, although moral therapy was generally linked to medical therapy in the total therapeutic programme, and although the non-restraint system was not rigidly observed in many asylums, these two themes were the most visible ones around which the nascent psychiatric profession emerged in early Victorian Britain.

3. THEORY AND PRACTICE IN THE NEW PROFESSION

British psychiatry acquired its professional trappings – professional organization and a specialized journal – in the late 1840s, at roughly the same time as equivalent events in France, Germany, and America.[14] The first meeting of the Association of Medical Officers of Asylums and Hospitals for the Insane (now the Royal Medico-Psychological Association) was held in 1841, under the stimulus of Samuel Hitch, resident physician to the Gloucestershire Asylum. For its first dozen years or so the Association remained precariously small, attracting attendances of only ten or twelve to its annual meetings, held each year in a different asylum, so that its members could compare the various therapeutic programmes employed.[15] By the time that the Association established its own journal in 1853 (*The Asylum Journal*, now the *Journal of Mental Science*), another periodical devoted to psychiatry had already been founded. This was *The Journal of Psychological Medicine*, which survived from 1846 to 1863 under the editorship of its promoter, Forbes Benignus Winslow (1810–1874) and was briefly reestablished between 1875 and 1883 by Winslow's son. Although there was an inevitable sense of rivalry between the two journals, they actually served the complementary functions which are indicated by their titles. For, during its early years, *The Asylum Journal* was largely concerned with the practical and professional matters involved in running the growing number of public asylums. Much journal space was devoted to analysing the annual reports of the Commissioners in Lunacy, the official body which oversaw the Victorian asylums; to publishing articles on asylum design, statistics, or therapeutic experience; and to providing British doctors with descriptions of asylum life and its problems in America and in Europe. In a sense, Winslow's journal was more intellectually ambitious, for it published rather more strictly clinical material, but it suffered from the lack of any formal professional support and probably from Winslow's own rather acerbic personality. Winslow was also the owner of two private mad-houses, and his journal naturally tended to support the private sector, or 'trade in lunacy.'

Nevertheless, the existence of these two journals by the 1850s attests to the extent of British medical interest in psychiatry; the demise of Winslow's journal also underscores the fact that the possible career structure in psychological medicine in Britain was not such as would permit the leisured and systematic investigation of serious mental disorder. Although the public asylum physicians achieved many of their aims – a network of compulsorily erected and publicly financed asylums, and the requirement of full-time resident medical practitioners within those asylums – they fell victims to their own limited success. Except for a few posts – such as Chief Medical Officer to the Privy Council, the Commissioners in Lunacy – medical careers in the public sector remained badly paid and low in prestige during the middle decades of nineteenth century Britain. Asylum physicians were grouped with Poor Law Medical Officers and Medical Officers of Health in running the portions of the Victorian medical service financed by the State.[16] The public asylums catered for a larger portion of the public than did the Poor Law Infirmaries, for many of those who would have been treated for their general medical problems in the Voluntary Hospitals, if diagnosed insane generally ended up in a county asylum. From around mid-century the private 'trade in lunacy' – the keeping of a licensed house with paying psychiatric patients – declined in importance relative to the county asylums.[17] This meant that there were gradually diminishing opportunities in the private sector, ambitions were thwarted by county officials anxious to keep asylum running costs to an absolute minimum, and by the silting up of asylums with chronic cases who were beyond hope of recovery and who lived monotonous, institutional existences for years. Unlike the part-time posts in general voluntary hospitals, which served as entrees into lucrative private practices, posts in insane asylums were full-time, and whilst advancement within the system could lead a young resident medical officer to the better-paying post of medical superintendent, the latter post was largely administrative, as average county asylum size increased from 116 patients in 1827 to 802 in 1890.[18] Success led to diminished clinical or scientific opportunities. Small wonder that recruitment of good people was difficult, or that ambitious young doctors like Henry Maudsley, or James Crichton-Browne used short term appointments in the asylums as opportunities to gain clinical and pathological experience before seeking more prestigious appointments in general hospitals, medical schools or higher government circles. Maudsley left the Manchester Royal Lunatic Asylum, Cheadle, after three years, becoming shortly afterwards Professor of Medical Jurisprudence at University College London; Crichton-Browne (1840–1938) spent nine productive years as Medical Superintendent to the West Riding Asylum (Yorkshire), before in 1875 becoming the Lord Chancellor's Visitor in Lunacy. He established the first formal pathology department in a British asylum while at the West Riding, but it was a relatively informal affair and his talents were recognized only by a part-time lectureship in the nearby medical school in Leeds. For Crichton-Browne as for other eminent Victorian psychiatrists, promotion was not through the academic ranks and meant a diminution in his clinical responsibilities.[19] Indeed, in mid-century, Thomas Laycock (1812–1876), who in addition to a chair in the practice of medicine in Edinburgh also held a lectureship in medical psychology there, came as close as anyone in Britain to devoting himself full-time to academic psychiatry. Though the London medical schools began appointing lecturers in mental diseases around the same time, these were part-time posts which were usually combined with private practice and the operation of a private madhouse. Laycock himself was a fertile thinker who first applied the reflex concept to cerebral functions and developed a sophisticated notion of the unconscious.[20] His approach to medical psychology was rather through neurology than psychiatry; consequently he belongs more appropriately to the very distinguished nineteenth century British neurological tradition, which also included such clinicians as John Hughlings Jackson, Sir David Ferrier, Sir William Gowers and Henry Charlton Bastian.[21] This tradition was never integrated into British psychiatry in the way that men like Griesinger and Wernicke in Germany attempted, with partial success, to create a genuine neuropsychiatry. One condition for this integration certainly existed in British asylums: diseases which we would nowadays classify as neurological – epilepsy, ataxias, Parkinson's disease, etc. – were common there.

The integration did not occur, however, and British psychiatry, though wedded to basic organic theories of insanity, remained rather circumspect in its approach to the diagnosis, classification, and treatment of insanity. Asylums became isolated institutions, cut off from the everyday world, and, too often, from mainstream medicine. Some indication of the difficulties facing nineteenth century British psychiatrists can be seen from the paucity of general, systematic works on the subject. There was not in Britain a tradition equivalent to that established by Esquirol, Guislain, or Morel in France, or Jacobi, Feuchtersleben and Griesinger in Germany, where the most eminent psychiatrists offered original and far-reaching surveys of the subject. In fact, two treatises (the second in multiple editions) served British alienists and general physicians as a source of systematic information on mental disorders for the half century following 1835. These were the *Treatise on Insanity* (1835) of James Cowles Prichard, and the *Manual of Psychological Medicine* (first edition, 1858, fourth edition 1879) of J. C. Bucknill and Daniel Hack Tuke. I should like to examine these works, for several of their common characteristics reflect broader aspects of British psychiatry during the period.

Prichard (1786–1848) was a Bristol physician of Quaker background who converted to evangelical Anglicanism as a young man. He remained devoutly pious and politically and medically conservative throughout his adult life. However, he was a man of vast erudtion who is still remembered for the anthropological and ethnological writings which culminated in the five-volumed, third edition of his *Researches into the Physical History of Mankind* (1836–47), a work which in its first edition (1813) contained original and influential views on heredity, geographical distribution of plants and animals, and the formation of human races. Prichard also published works on mythology and philology, on the vital principle, and a number of shorter pieces on medical topics such as fevers. His interest in psychiatric matters stemmed from early in his career (1811) when he had been elected physician to St. Peter's Hospital, a Bristol hospital for paupers which from early in the eighteenth century had admitted a high proportion of insane patients. In addition to the general psychiatric volume of 1835, Prichard wrote a *Treatise on Diseases of the Nervous System* (1822), dealing with convulsive and maniacal disorders, and, late in his life, a short work on the relation of insanity to jurisprudence (1842). This gave him the opportunity to expound the practical consequences of what was the most novel element of his 1835 *Treatise*, the concept of 'moral insanity'. Prichard left Bristol for London in 1845, when he has appointed one of the Commissioners in Lunacy and it was during his pursuit of these duties that he contracted the illness which led to his premature death in 1848.[22]

By contrast, Bucknill and Hack Tuke were both full-time psychiatrists. Bucknill (1817–1897) had a distinguished student career at University College London, and was contemplating a career in surgery when his health broke down and he moved to the warmer climate of south-west England. He consequently became medical superintendent of the Devon County Asylum at Exminster (1844 to 1862). It was there that he established his name, as first editor of *The Asylum Journal* (1853 to 1862), and as co-author, with Hack Tuke, of the *Manual of Psychological Medicine*. Bucknill left asylum life in 1862, to become Lord Chancellor's Medical Visitor of Lunatics, from which post he retired into private practice in 1876. In 1878 he founded, with Hughlings Jackson, David Ferrier, and J. Crichton-Browne, *Brain: a journal of neurology*, in itself a reflection of developments in the neurosciences in Britain.

Daniel Hack Tuke (1827–1895) was the great-grandson of the founder of the York Retreat, and for several years was visiting physician to that institution. He eventually settled in London, where he combined a private practice with a lectureship on mental diseases at Charing Cross Hospital, and a long-term association with Bethlem Hospital (Bedlam), the famous London lunatic establishmet. His *magnum opus*, still a work of considerable historical value, was the *Dictionary of Psychological Medicine* (2 vols., 1892), which was probably as close as British alienists ever came to a work conceived and executed on the generous scale so common in Germany.

The first edition of Bucknill and Tuke's *Manual* was separated from Prichard's *Treatise* by twenty three years, and the works naturally exhibit considerable differences, as do the first and last editions of the *Manual*. Beneath the differences, some the result of accumulation of empirical information, others of shifting fads in regimen or specific new theories about the cause, diagnosis, or prognosis of insanity, lay some striking continuities of approach and style. Five of these are particularly worth stressing, for they reflect more general characteristics of British psychiatry in the middle decades of the century.

3.1. *Although Operating Within an Explicitly Psychosomatic Framework, the Ultimate Commitment Was Always to an Organic Idea of Insanity*

There were some exceptions, but nineteenth century British psychiatrists had difficulty accepting a notion of *primary mental disease*: some of the difficulty was theological, for the equation of mind with soul protected the latter and hence the former from the ravages of disease and death. As I have suggested elsewhere, there were also strategic professional motives at stake, for the claims of medical men against clergymen or lay reformers as the primary experts in insanity relied on the notion of organic disease for much of its validity.[23] Nevertheless, there were problems with an organic model. Prichard, for instance, combated the phrenological doctrines of Gall and Spurzheim particularly for what he saw as phrenology's inherent materialism, and advanced instead a notion of a unified and indivisible mental faculty using as its instrument a unified cerebrum which consequently could not be localized as subserving discrete mental functions. Against this backdrop, however, he insisted on the idea of partial insanity – the moral sense could be diseased without disturbance of the intellectual faculties. Furthermore, though his underlying dualistic philosophy of mind could not easily accommodate it, he allowed for the primary efficacy of moral therapy. Prichard was the victim of the difficulties created by Descartes when he divided the world into two incommensurate categories, mind and matter.[24]

These tensions were less acute in Bucknill and Tuke, partly because they simply set aside the metaphysical question of *how* mind and body acted on each other and concentrated on the pragmatic fact that minds and brains are found together. In 1853 Bucknill had insisted that the distinction between organic and functional diseases is spurious: *All diseases are organic*, he wrote, even if we are unable to discover the underlying pathological changes.[25] In the *Manual*, he and Hack Tuke summarized their position as follows:

> The brain, like every other organ of the body, for the performance of its functions, requires the perfect condition of its organization, and its freedom from all pathological states whatever. Consequently, the existence of any pathological state in the organ of the mind will interrupt the functions of that organ, and produce a greater or less amount of disease of the mind – that is of insanity.[26]

Elsewhere, they remarked that since even perceptions and sensations must result in some minute change in the nervous system, there was no theoretical reason why moral therapy should not be effective, even if the disease were organic.

3.2. *Nosologies Were Relatively Simple and Based on Behavioural Characteristics*

Although the Greeks had provided a basic vocabulary – mania, melancholia, dementia – for classifying mental disorders, late eighteenth and early nineteenth century British nosologists such as William Cullen, Thomas Arnold, and John Mason Good had produced rather clumsy and elaborate schemes. Pinel, however, had returned to the basic simplicity of classical authors and, with some exceptions, nineteenth century British authors had been content to work within the Pinelian framework as modified by Esquirol. Thus Prichard divided insanity into two grand forms, moral or intellectual, with the latter sub-divided into monomania (with a frequent element of melancholia), mania, and incoherence or dementia.[27] Bucknill and Tuke added a third general class, those involving the propensities or passions, though in practice they preferred the simple classification of idiocy; dementia (primary or secondary); delusional insanity (either manic or depressive); emotional insanity (either 'moral' insanity or melancholy without delusion); and mania, either acute or chronic.[28] They recognized that epilepsy and general paresis could complicate any of the above diagnoses, but believed that these latter conditions did not warrant primary diagnostic categories of insanity in their own right. Although they continued to use this same classification through the final edition of their *Manual*, by 1879 they were aware of the desirability of an *aetiological* classification.[29] Only one good analogy seemed worth considering, though, and this was the relationship between intoxication and insanity. Accordingly, they suggested that toxic factors as yet undetermined might eventually be implicated in the causation of the various forms of insanity. Until these were identified, however, they stressed that speculation was of little use.

3.3. *The Organic Commitment Led to a Search for Pathological or Patho-physiological Mechanisms to Explain Symptoms*

The general medicine which was developing in the early nineteenth century has been called 'hospital medicine' by Ackerknecht. It derived largely from the Paris hospitals after the re-founding of the French medical schools in 1794, and was based on the notion of *local* pathology, the practice of careful physical diagnosis, the systematic use of autopsies to correlate clinical signs and symptoms with pathological lesions, and the use of large series of cases numerically reported to establish firmer diagnostic and therapeutic indications.[30] Pinel was an internist as well as a psychiatrist, and certainly the psychiatry of Esquirol, Georget, Foville, Calmeil and other French doctors reflected many features of this hospital medicine. Autopsies were more routinely performed in French asylums than British ones but British alienists were aware that various attempts had been made to explain the symptoms of insanity in terms of the routine categories of 'physical' diseases such as tuberculosis or cirrhosis. They were also aware that such attempts were generally indecisive and mutually contradictory. Nevertheless, Prichard discussed at considerable length French 'patho-psychiatric' work, and while favouring explanations which involved either local hyperaemia and inflammation, or the sympathic neurological response to thoracic or abdominal inflammation, he realized that definitive patho-physiological explanations had not yet been produced.[31] Indeed, on occasion Prichard seemed genuinely relieved by the failure of pathology, for it seemed to support his belief in the separate existence of mind from brain. Likewise, Bucknill and Tuke recognized that many cases of chronic insanity had been autopsied without uncovering any structural defects in the brain and central nervous system. To explain this apparent anomaly, they developed an elaborate notion of nerve 'force', normally generated by the healthy brain but under conditions of local vascular change unable to exercise its 'normal' functions. This was ultimately a nutritional problem, but once set in motion could lead to compensatory mechanisms in other parts of the brain so that the relative balance was lost and chronic symptoms without visible structural changes could occur.

Now, we recognize explanations of the kind put forward by Tuke and Bucknill as essentially speculative, based at best on analogy but with little in the way of specific or direct evidence to support them. Yet Bucknill and Tuke shared a horror of mixing overt metaphysics with their psychiatric writings and placed their own work firmly within the pragmatic, empirical British tradition. This was easier because they eschewed a new or esoteric vocabulary and based their pathophysiological discussions on what they conceived to be the sound work of men like Rokitansky and Virchow. In this way, the conclusion to a fifty-page section on the pathology of insanity can end with the following summary which, while admirably clear in its expression conveys little in the way of information:

> The theory of partial insanity, without appreciable change of the brain, is as follows: – When the disease first exists, it is attended by pathological states of the cerebral vessels. A morbid condition of the cerebral organization is occasioned, attended by the phenomena of insanity. After a short time, the vessels recover their tone, the brain is nourished, and its size maintained as a whole. But the original balance of its organs is not

regained; their nutrition having been impressed in the type or mould of their diseased state. Perhaps some of the cerebral organs encroach on others by their actual bulk; undoubtedly, some of them overbear others by their greater activity. The result is chronic mental disease, of a nature which leaves behind no pathological appearance.[32]

'Brain mythology' was not a German monopoly.

3.4. *The Organic Commitment Was Accompanied by a Neglect of Normal Psychology or Even Neurophysiology*

Until the 1870s, when Francis Galton (1822–1911) began elucidating his theories of psychological functions based on the notion of the faculty, much formal psychology in Britain can be seen as a continuation of the work started by John Locke (1632–1704). Locke stated that at birth, the mind is a blank tablet (*tabula rasa*) on which impressions are made through sensations and the combination of these sensations into reflections. Experience was thus the source of all knowledge. In the eighteenth century, various attempts were made by men such as John Gay (1699–1745), David Hartley (1705–1757) and Joseph Priestley (1733–1804) to develop the *association of ideas* as the mechanism through which the mind works. Locke had, in a rather offhand comment, remarked that madmen reason correctly from false premises, and a number of eighteenth century writers on insanity used this starting point, together with the association of ideas, to explain something of the aetiology of madness and the mental content of the insane mind.[33] In the nineteenth century this psychological tradition was continued by James Mill (1773–1836), his more famous son, John Stuart Mill (1806–1873), and, in an evolutionary context, Herbert Spencer (1820–1903) and Alexander Bain (1818–1903).[34]

Curiously enough, neither Prichard nor Hack Tuke and Bucknill made use of this or any other psychological tradition. Prichard began his *Treatise* with preliminary remarks on the definition and nosography of insanity; Hack Tuke and Bucknill their *Manual* with a history of the treatment of insanity. Their concern with 'diseased' minds takes no cognizance of 'normal' ones, a rather striking omission particularly when one recalls how much German-speaking psychiatrists like Feuchtersleben and Griesinger made use of theories of normal psychological function. In the case of Bucknill and Tuke, I suspect the reason for this omission lies in their belief that psychology was too 'metaphysical' and introspective, and they were all too keen to establish psychiatry on a firm empirical basis. Even so, the loss was considerable, for their case histories lack subtlety when it comes to discussing what they call 'the mental state' of insane patients. For instance, the history, mental state and physical condition of one patient, described as a case of 'acute mania subsiding into quiet melancholia' was given as follows:

An engineer; a clever, industrious man, of steady habits. Three months before admission experienced a severe disappointment, in not getting an order for a certain steam-engine which he had calculated upon, he became excited and irritable in manner, neglected his work, and acute mania gradually came on. *Mental State.* – Extreme excitement; believes that he is going to be shot; asks everyone why he is not killed, and begs of them to kill him, shouting all night long; tears his clothes, destroys his bedding, scribbles on the walls and doors; jumps at the gas-pipes, and attempts to pull them down; very destructive and violent; wets and dirties his bed; miscalls persons, fancying he has seen everyone before: no power of fixing his attention. *Bodily Condition.* – Expression pale, wild, haggard; skin clammy, extremities cold, head cool; losing flesh; pulse small and quick, bowels constipated.[35]

Histories like this do not satisfy the modern reader.

If Prichard, Hack Tuke and Bucknill did not seem too interested in the nuances of either the normal or diseased mind, neither were they interested in integrating another rich nineteenth century British field of investigation: reflex physiology. By mid century, W. B. Carpenter (1813–1885) and Thomas Laycock were extending the notion of the reflex arc to the higher cerebral centres, and both Bain and Spencer drew on this work in their evolutionary psychologies, as did Hughlings Jackson in his neurological writings. Attempts by men like Meynert in Vienna to apply reflex physiology to psychiatry were not satisfactory in the long run, but Bucknill and Tuke did not even bother to consider the possibility to dismiss it, a reflection no doubt of their general neglect of the 'normal'.

3.5. *There Was Virtually No Concern with the 'Neuroses'*

In two interesting but too little known monographs, Professor Lopez-Piñero and his colleague have shown how the concept of 'neurosis' originally was developed by general physicians such as William Cullen (1710–1790), John Brown (1735–1788) and other late eighteenth century figures.[36] It was only from the middle of the nineteenth century that a modern notion of neurosis began to be incorporated into psychiatric thought, through, as Fischer-Homberger has shown, the idea of 'traumatic-neuroses' which was much discussed from the 1860s.[37] While there was in Britain a considerable interest in phenomena such as mesmerism (the word 'hypnosis' was actually coined by a Manchester surgeon named James Braid), animal magnetism, somnambulism, ecstatic states, and, above all, hysteria, these phenomena were by and large not of much concern to nineteenth century British psychiatrists. Prichard to be sure included a final chapter in his *Treatise* which reviewed the history of what he called 'animal magnetism' and described interesting cases of somnambulism and maniacal ecstasy,[38] but one finds very little of this kind of material in the pages of *The Asylum Journal*, and hysteria rates less than two pages out of 556 in the first edition of Tuke and Bucknill's *Manual*. Much more work needs to be done on the place of these phenomena in Victorian medicine and society before we can fully understanding these matters, but the extent to which British psychiatry revolved around institutional treatment meant that what, despite Bucknill, were called the 'functional' nervous disorders were much more likely to be seen by general physicians, gynaecologists, and neurologists than by alienists.[39]

4. CONCLUDING SUMMARY

In this paper I have attempted to sketch briefly the major formative social forces and the principal intellectual and practical themes which are central to nineteenth century British psychiatry. Any one of the characteristics which we have looked at in conjunction with the systematic treatises of Prichard and Bucknill and Tuke would bear examination in greater detail, since in actual fact there was rather less unanimity of opinion than my brief remarks may have suggested. Nevertheless, the most striking feature – the identification of the British psychiatric profession with the asylums – will stand.

By the 1870s, there were signs of change, although asylums continued to be the dominant reality in British psychiatry until after World War I. The changes were partly catalysed by the collapse of the optimism which had generated the earlier reforms, for with the gradual silting up of the institutions with chronic cases, and the inevitable mechanization and regimentation which accompanied their increase in size ('A gigantic asylum is a gigantic evil,' wrote one psychiatrist), it became more frustrating for psychiatrists to throw creative energy into the asylums. We have already seen some indications of the resulting new directions – the establishment of a department of pathology and research laboratories in the West Riding Asylum, the foundation of a new journal, *Brain*, in which psychiatrists and neurologists joined forces, and new attempts by men such as Henry Maudsley to broaden the basis of psychiatric thought. Maudsley (1835–1918) is today remembered primarily because of the psychiatric hospital which he founded late in his life. From the late 1860s, however, he produced a series of popular and scholarly monographs which made him the first British psychiatrist since Conolly to acquire an international reputation. His most important work was *The Physiology and Pathology of the Mind* (1st ed. 1869), a monograph in which he attempted to integrate psychology, reflex physiology, and psychiatry into a single synthetic whole.

Apart from the addresses of Sir Aubrey Lewis and Dr. Walk, Maudsley remains a too little appreciated figure.[40] Nevertheless, his career and writings highlight a number of features of late Victorian and Edwardian psychiatry: the enrichment of psychiatry by reference to 'normal' psychology; the slow development of a psychiatric profession external to asylum life; the attempt to apply the 'lessons' of psychiatry to the problems of everyday life; and the quiet pessimism which undermined the rhetoric of evolutionary progress. That his writings also bear the explicit stamp of his moral values and ethical judgements is not so much a comment on late Victorian psychiatry as on the nature of medicine and the processes which generate concepts of health and disease.

University College London

NOTES

1 E. H. Ackerknecht, *A Short History of Psychiatry*, 2nd ed. (New York, 1968); Karl Jaspers, *General Psychopathology*, trans. from 7th German ed. (Manchester, 1963); Henri Ellenberger, *The Discovery of the Unconscious* (New York, 1970).

2 E. P. Thompson, in a brilliant essay entitled 'The peculiarities of the English' has defended English empiricism; in his collection, *The Poverty of Theory* (London, 1979).

3 E.g. Gregory Zilboorg, *A History of Medical Psychology* (New York, 1941).

4 For discussions of 'moral therapy', see E. T. Carlson and Norman Dain, 'The psychotherapy that was moral treatment', *Amer. J. Psychiatry* 117 (1960): 519–524; K. M. Grange, 'Pinel and eighteenth century psychiatry', *Bull. Hist. Med.* 35 (1961): 442–453.

5 Quoted in Andrew T. Scull, *Museums of Madness: The Social Organization of Insanity in 19th Century England* (London, 1979), p. 133.

6 Michel Foucault, *Madness and Civilization*, trans. from the French (London, 1971).

7 On therapeutic optimism, see Ida Macalpine and Richard Hunter, *George III and the Mad Business* (London, 1969).

8 P. Pinel, *A Treatise on Insanity* (Sheffield, 1806), reprinted (New York, 1962) with an introduction by Paul Cranefield, Section 5, Scull (note 5); David Rothman, *The Discovery of the Asylum* (Boston, 1971).

9 On partial insanity, see Henry Werlinder, *Psychopathy: A History of Concepts* (Uppsala, 1978).

10 Samuel Tuke, *Description of the Retreat* (York, 1813), reprinted (London, 1964), with an introduction by R. Hunter and I. Macalpine.

11 Scull (note 5), Chapters 1, 2; Kathleen Jones, *A History of the Mental Health Services* (London, 1972).
12 John Conolly, *Treatment of Insane Without Mechanical Restraints* (London, 1856), reprinted (London, 1973), with an introduction by R. Hunter and I. Macalpine. The passage quoted is from Hunter and Macalpine, *Three Hundred Years of Psychiatry 1535–1860* (London, 1963), pp. 1034–1035.
13 J. C. Bucknill in *The Asylum Journal* 1 (1855), p. 16.
14 For comparative events, see Gerald Grob, *Mental Institutions in America* (New York, 1972), Chapter IV; J. G. Howells (ed.), *World History of Psychiatry*, Chapters 4. 10, and 19.
15 Alexander Walk and D. L. Walker, 'Gloucester and the beginnings of the R. M. P. A.', *J. Ment. Sci.* 107 (1961): 603–632.
16 For general discussions, see Ruth Hodgkinson, *The Origins of the National Health Service* (London, 1967).
17 W. Parry-Jones, *The Trade in Lunacy* (London, 1972).
18 Scull (note 5), p. 198.
19 Henry Viets, 'West Riding, 1871–1876', *Bull. Hist. Med.* 6 (1938): 477–87.
20 Roger Smith, 'The background of physiological psychology in natural philosophy', *Hist. Sci.* 11 (1973): 75–123.
21 For some of this work, see R. M. Young, *Mind, Brain and Adaptation in the Nineteenth Century* (Oxford, 1970), Chapters 6–8; Arthur M. Lassek, *The Unique Legacy of Doctor Hughlings Jackson* (Springfield, Ill., 1970). On neurological disease in Victorian asylums, see R. Hunter and I. Macalpine, *Psychiatry for the Poor* (London, 1974).
22 For a good biographical account, see George Stocking, Jr., Introduction to a reprint of Prichard's 1813 *Researches into the Physical History of Man* (Chicago, 1973).
23 W. F. Bynum, 'Rationales for therapy in British psychiatry, 1785–1830', *Med. Hist.* 18 (1974): 317–334; and Scull (note 5).
24 W. Bynum, 'Varieties of Cartesian experience in early nineteenth century neurophysiology', in S. Spicker and H. T. Engelhardt, eds., *Philosophical Dimensions of the Neuro-medical Sciences* (Dordrecht, 1976), pp. 15–33.
25 J. C. Bucknill, in *Asylum Journal* 1 (1855), p. 77.
26 J. C. Bucknill and D. H. Tuke, *A Manual of Psychological Medicine* (London, 1858), p. 353.
27 J. C. Prichard, *A Treatise on Insanity* (London, 1835), Chapter 1.
28 Bucknill and Tuke (note 26), pp. 86–91.
29 Bucknill and Tuke, *Manual*, 4th ed. (London, 1879), p. 310.
30 E. H. Ackerknecht, *Medicine at the Paris Hospital, 1794–1846* (Baltimore, 1967); R. H. Shryock, *The Development of Modern Medicine* (London, 1948), Chapters 9–10.
31 Prichard (note 27), Chapter 5.
32 Bucknill and Tuke (note 26), pp. 401–402.
33 Robert Hoeldtke, 'The history of associationism and British medical psychology', *Med. Hist.* 11 (1967): 46–65.
34 In addition to Young (note 21) and Smith (note 20), see R. M. Young, 'Association of ideas', in P. P. Wiener (ed.), *Dictionary of the History of Ideas* (New York, 1973).
35 Bucknill and Tuke (note 26), p. 536.
36 José Lopez-Piñero, *Origenes históricos del concepto de neurosis* (Valencia, 1963); and José Lopez-Piñero and José Morales Meseguer, *Neurosis y psicotherapia* (Madrid, 1970).

[37] E. Fischer-Homberger, *Die traumatische Neurosen* (Bern, 1975).
[38] Prichard (note 27), Chapter 12.
[39] Fred Kaplan, *Dickens and Mesmerism: The Hidden Springs of Fiction* (Princeton, 1975); Fred Kaplan, 'The Mesmeric Mania: the early Victorians and animal magnetism', *J. Hist. Ideas* **35** (1974): 691–702.
[40] Aubrey Lewis, 'Henry Maudsley: his work and influence', in his *The State of Psychiatry* (London, 1967); Alexander Walk, 'Medico-Psychologists, Maudsley and the Maudsley', *Brit. J. Psychiat.* **128** (1974): 19–30.

An earlier version of this paper was published in *History of Psychiatry. Proceedings of the 4th International Symposium on the Comparative History of Medicine - East and West*, Tokyo, Taniguchi Foundation, 1982.

STEFANO POGGI

COMMENTS ON W. F. BYNUM, *THEMES IN BRITISH PSYCHIATRY, J. C. PRICHARD (1785–1848) TO HENRY MAUDSLEY (1835–1918)*

(1) Bynum's paper takes us through the important moments in the institutional history of British Psychiatry from 1830 until about 1870. He also makes a suggestion: that we should explore certain connections not only with the psychological debate that was going on at the same time in Europe, but also with the 'influential metaphysics' by which this debate was inspired.

If we take up this suggestion, then the first major problem we encounter is the influence of Phrenology. There is no need to dwell on the influence it has had: Bynum himself reminds us and the matter has been amply discussed in the last ten years by R. M. Young, R. Smith, Di Giustino, Cantor, Shapin and Cooter.[1] These studies have shown that Phrenology was influenced in several different ways by the whole psychological debate that went on in Great Britain from 1820 to about 1850. This influence was most sharp in the field of psychiatric conceptions – for example the theory of the organic character of mental illness and of the 'moral therapy' that, it was claimed, could be used to treat it. In such a theory the brain is seen as the organ of the mind. Certain of the functions of the brain can be impaired and this can prevent the mind from functioning fully and can thus cause "partial insanity". The mind itself, however, in so far as it is an immaterial entity, can still react to 'moral therapy' capable of releasing all its potentialities.

In reality, as Bynum reminds us, this theory is influenced not only by Phrenology, but also – and despite the obvious contrast between the two concepts – by the psychology of the faculties, the traditional psychology, that is, of the Universities. Clearly then, the specific characteristics of British Psychiatry are better grasped if we keep in mind the extent to which it shares the clear distinction which is the fundamental principle of psychological analysis of the 'normal' functions of the mind.

Until the 1860's British Psychology, as is well known, presents a distinct Cartesian character. It starts by distinguishing clearly between the '*res extensa*' and the '*res cogitans*' and then concentrates on the latter.[2] Thus one gets a 'science of mind' that numbers some decisively philosophical elements and that is based on the guiding concept of the associationist tradition.

And it is precisely on the basis of the fruitfulness of the associationistic approach that the claim to the British lead in having established psychology

Michael Ruse (ed.), Nature Animated, 243–266.

is made by J. St. Mill. It is worth dwelling on this point a while in order to expand it further.

(2) As J. St. Mill stated in his review of *The Senses and the Intellect* by A. Bain, "the sceptre of psychology has decidely returned to this island. The scientific study of mind, which for two generations, in many other respects distinguished for intellectual activity, had while brilliantly cultivated elsewhere, been neglected by our countrymen, is now nowhere prosecuted with so much vigour and success as in Great Britain".[3]

In Mill's opinion, Bain's work represents the most complete statement of associationism in the British philosophical tradition. Bain's definition of the law of association as a 'governing principle' means the 'laws of mind' can be simplified to the maximum and facts connected with those laws can be generalised as far as possible.[4] Mill, indeed, starts his analysis of Bain's theory by stating the basic characteristics of associationism and on this point he remains faithful to the theory he had already set out in his *System of Logic* of 1843. This maintained that there was no purpose served in separating out into their ultimate parts mental operations concerned with every logical inference.[5] At the same time, however, Mill recognises the need "to endeavour to ascertain the material conditions of our mental operations".[6] Although later to be left to one side in his *System*, this idea originally came from a concern to delineate the different domains of logic, psychology and metaphysics. It was further adopted by Bain and was held by Mill who, for his part, clearly stated that the accusation of 'materialism' levelled at this method of "interpreting the phenomena of mind" was absolutely unfounded. If this accusation had had any basis at all," all theories of the mind which have any pretension to comprehensiveness must be materialistic".[7]

Mill therefore recognises on the one hand that the way in which the brain is the 'instrument' of thought is destined to remain 'mysterious', but on the other hand he points out that "many indisputable pathological facts" prove a connection exists between the brain and thought. So Bain is right to emphasize that the dynamic nature of mental life is connected to the spontaneity of the brain's activity: in fact, the brain is 'a self-acting instrument'.[8]

Psychology – a science which, as Mill had stated in his *System of Logic,*[9] is concerned with "the laws, whether ultimate or derivative, according to which one mental state succeeds another" – received also a coherent definition from Bain. Its strong point was that it established the connection between the associationistic laws of the mind and a physiological analysis of sensations. This connection was established by the "law of relativity", by means of which, according to Bain, "we are never conscious at all without experiencing transition or change" and we register this change with our 'sensory'. Any change in our sensations gives us the opportunity to contrast one with another, to establish this "comparison" and then to discover its 'discrimination'; this 'discrimination' is, according to Bain, the "basis of our intelligence" and the "commencement of our knowledge".[10]

Basically then, this 'law of relativity' sets itself up as a basic premise for any psychological and physiological analysis, as a law regulating the 'states of mind' as soon as they manifest themselves in the sensations, in the 'feelings'.[11]

When Mill attributes Britain's regained lead in the 'scientific study of mind' for the most part to Bain, he also shows his awareness of the changes in British philosophy that had been prompted by the physiological discoveries of the time. The work of Ch. Bell. C. Ludwig, E. H. Weber gave rise in fact to a particularity lively debate one, which also tackled the problem of materialism.[12] This problem was to become gradually more substantial with the emergence of reductionist theories deriving from a new approach to the problems of the 'mind', its 'states' and its 'laws'.[13]

Mill's conclusions about Britain's lead in psychology seem to be fully confirmed by the fact that as early as 1855 H. Spencer, who Mill considered a less 'sober' thinker than Bain, had published *The Principles of Psychology.*[14] Spencer's work – still earlier than Bain's – sets out a uniform and systematic approach to the problem of the mind's structure, an approach that is formulated from an evolutionistic standpoint. But the fact that Spencer, too, adopted the idea of 'comparison-discrimination' between the various states of mind as states of consciousness is significant.[15] He points therefore how the change in sensations and the consciousness coincide, following the traditional procedures of associationism which J. St. Mill had earlier set out so clearly in the introductory pages of the *System of Logic.*[16]

(3) Assuming that the study of British Psychiatry should not be confined to the mere history of the institutions and of the social and ideological groups that are behind the practise of non-restraint therapy, it will clearly be useful to compare the concepts of Psychiatry with those of the science of mind. The need for such a comparison seems to be emphasized by the results that Bynum has obtained from examining the two principal treatises of British Psychiatry – that of Prichard, and that of Bucknill and Tuke. As Bynum shows, both Prichard and Bucknill and Tuke reveal considerable uncertainty in formulating the idea of mental illness. Furthermore, despite

explicit allusion to the pathophysiology of Rokitanski and Virchow and their declared dislike for mingling metaphysics with psychiatric discussion, their studies fail to explore the connections that might exist between Psychiatry on the one hand and the psychology of the 'normal mind' and neurophysiology on the other. This implies that they accept the Cartesian separation of 'mind' and 'brain'. On the one hand, as Bynum reminds us, their arguments belong to 'brain mythology', while on the other (perhaps even when, as in the case of Bucknill and Tuke, a pragmatic conception of the 'mind-brain' link prevails) they leave the 'science of mind' to play the principal role in the analysis of the so-called 'mental states'.

If we consider the methodological approach of British Psychiatry at least until the end of the sixties (and I think that full and careful checks are required on the basis of the suggestions made by Bynum), we find that it shares to a considerable extent the non-reductionist view of the mind.

At the same time it seems quite clear that British Psychiatry in those years was unable to take proper account of the progress that was being made, admittedly with great difficulty, in British Psychology, in the work of Bain (Spencer constitutes a case apart) and of physiologists as Carpenter and Laycock.[17]

Since it lost all faith in Phrenology in the 1850's, British Psychiatry remained remote both from the evidence derived from the physiology of the nervous system and from the psychological controversy which this last was largely responsible for provoking. British Psychiatry may well admit the need to explain the symptoms of mental illness on the basis of pathophysiology, but the only thing it actually does, as Bynum points out, is to introduce the vague notion of 'nerve force'. This notion is in fact present also in Carpenter and can be linked to Bain's not always convincing ideas about 'nerve substance' and 'nerve quality'. At the same time, however, it most obviously recalls Cullen's theories of magnetism.[18]

(4) Although it fails to concentrate specifically on a psychological and physiological investigation of the 'normal mind', British Psychiatry of the years 1830 to about 1860 was patently inspired by a type of 'influential metaphysics' that was not substantially different from the one that was a central feature of the whole British psychological debate in the last century until the sixties.

I would like to turn now from Britain to France and Germany to consider what happened there in the same years.

In the first place we note that psychological and physiological thinking in France from Cabanis and Maine de Biran (the latter a committed supporter of the physical aetiology of mental illness) displays a Cartesian division analogous to the one present in psychological (and thus inevitably psychiatric) thinking in Great Britain. In France, up until the Seventies, only scientists took any real interest in establishing an experimental science of psychic phenomena. This means that the French, in their discussion of the problem (and what they have to say is in harmony with hospital medicine) tend to give decisive importance to the autoptic examination of the brain. At this same time, and while Flourens was conducting his experiments and attacking Phrenology, French physiologists were following the lead of Magendie and moving closer to a determinist view of the processes of living organisms. Such a view was to gain a certain degree of confirmation from the admittedly problematical work of Bernard. Here too belong the psychiatric theories of Morel and the spread of the theory of heredity also in literature.[19]

On a more specifically psychological level, however, we must not overlook the work of Th. Ribot, where the Cartesian division between "mind" and 'brain' was rendered somewhat problematical under the influence of, first of all, the psychological debate in Britain and later of the debate in Germany.

In his *La psychologie anglaise contemporaine*, Ribot shows a very marked interest in the work of Spencer.[20] Spencer, in fact, as Ribot stressed, pointed to and set out the basic guidelines for integrating and systematically organising the connections between "states of mind" and physiological processes. What Ribot so wholeheartedly admires then is Spencer's general philosophical conception, one which emerged as early as the first edition (1855) of the *Principles of Psychology*. Seeing that this is founded on the idea of the "unknowable", Ribot seems to parallel certain of Spencer's ideas with those of Schopenhauer's. Consequently, Spencer's conception lends itself to a definition of psychology, its objectives and domain, as the science of mental (psychic) phenomena such as they are, that is, not investigating either their origins or their purpose.[21]

On this basis, Ribot therefore delimits the objectives of psychology in terms that show him to be extremely wary of adopting any "reductionist" hypothesis. Thus he takes up a position opposing those ideas characteristic of Comtian philosophy, which where still held in opposition to J. St. Mill by E. Littré in 1866, who used arguments aimed at stressing the need not to overlook 'cerebral physiology' and no to reduce psychology to the study of 'ideology and logic'.[22]

In Ribot's opinion, the psychology of J. Mill, J. St. Mill, Bain, Lewes, Maudsley and above all Spencer achieves a balance between laws of thought (or of the 'mind') and "analysis of sensations" which enables psychology to be established as a science and prevents it from becoming some metaphysical discipline of either materialistic or spiritualistic nature.

In his analysis, Ribot therefore points out that British psychology:

- isolates the data (it obviously does not dismiss the physiological approach but it does not link it with any materialistic generalizations);
- defines the laws that connect these data;
- justifies at the same time, by determining the criteria and methodology, the use of 'internal experience' or 'introspection'.[23]

The psychological debate in France in the Sixties and Seventies was divided between on the one hand, the 'reductionism' of Comte's school of thought, which moreover had been supported by no less an authority as Laplace, convinced as he was that psychology could be nothing other than an extension of physiology,[24] and on the other hand, the approach incorporating the "sens intime" which Maine de Biran's school appealed to. The French debate also touched on themes that originated in the debate of those '*idéologues*' close to Scottish philosophy. It was through Ribot, however, that the French discovered a term of reference in British psychology that was of crucial importance.[25] Thus it was that in the course of these years French culture as a whole came to be more open to the general philosophical vision associated with the British approach to the "scientific study of the mind". The rivality between the system of Comte and Mill's (and Whewell's) methods was in effect coming to an end, with Mill emerging victorious,[26] although we must still duly keep in mind the arguments Littré continued to use in defence of some of Comte's theories.[27]

Another essential work, in addition to Ribot's, which evidences the interest French culture took in the psychological and methodological debate in Britain is *De l'intelligence* by H. Taine, also published in 1870.[28] Taine, too, however, refers quite explicitly to many definitions taken from Condillac's sensism and on these lines but still following a general Lockian approach identifies the terms the British debate had used to develop a conceptual framework that in actual fact was not very different from that of the '*idéologues*'.[29] The balance between the laws of the mind and the analysis of the sensations achieved within British psychology is echoed quite importantly in Taine's work, who nevertheless, unlike Ribot, does not pay particular attention to Spencer but associates himself above all with J. St. Mill.[30] To be more exact, Taine adheres to the traditional ideas of associationism and states – evidently in connection with the law of relativity as defined by Bain – that the operations of intelligence are based on a double structure ("*la couple*") that always connects the distinguishing features of two sensations or of two ideas.[31] Furthermore Taine shows himself to be close to Tyndall's theories in the matter of what role 'imagination' plays in the growth of knowledge and he maintains that the normal state of our '*esprit*' consists of a series of '*hallucinations*' that are never brought to a conclusion. They reveal the dynamic nature of mental activity and represent, according to Taine, "the plan itself of our mental life".[32]

In addition we must not overlook the importance Taine's work places on the whole problema of the signs of natural processes, signs which the mind has to interpret, given that, in every case, what constitutes our 'original language' is the 'moral' evidence of the '*esprit*'.[33]

Ribot's work on the one hand and Taine's on the other evidence French awareness of the important position British psychology occupied in their country, one which, nevertheless, made and continues to make significant contributions to psychology particularly with Magendie, Florens and Bernard.

(5) Despite the fact that the French situation still poses a great many problems, which, except for the already complex and intricate question of the spread of Darwinism,[34] have hardly been investigated, there is however no doubt whatsoever as to the stimulus and direction British psychological analysis gave to the philosophical and to some extent the scientific debate in France. What in some aspects appears an absolute state of dependance on the British cannot however deny French culture the coherence and quality of its contribution not only to general physiology[35] but more specifically to neurophysiology,[36] particularly in the matter of distinguishing between motor and sensory nerves and the question of cerebral localisation.

The central core of the 'analysis of sensations' carried out by French science, after Magendie, is, however, the debate around cerebral localisation of psychic functions and in particular the problem of so-called 'aphasia'. The principal names involved in this debate were P. Broca, J. B. Bouillaud, J. Lordat, J. Baillarger, and A. Trousseau.[37]

Within this group serious differences developed between those who believed in an 'internal sense' and tended to interpret scientific data as proving the absence of any real link between laws of thought and their physical vehicle, as against those inclined towards a profound materialism and who adopted an essentially 'reductionist' standpoint.[38]

The most significant position in this debate was taken by P. Broca.[39] He gave the 'reductionist' interpretation of localisation his full support but he tempered this by recognising at the same time the extremely complex nature of the functional connection between brain (termed a 'logical machine' by Baillarger) and its faculties, especially those associated with spoken language.[40]

Although it is impossible to set up a connection between cerebral lesions and the failure of the 'faculties' to operate, it is nevertheless also true that pathological observations compel one to consider the existence of a more general and 'deeper' connection between the faculty of language and the entire range of the so-called intellectual faculties. The study of cerebral convolutions and the cortex may well provide a complete and systematic solution to the problem of localising any kind of 'faculty' or 'superior psychic function'.[41]

Thus the theory that gradually dominates is that aphasia represents a disturbance in the functions of 'intelligence' in general (that is intelligence as Taine was to define it in his *De l'intelligence*)[42] or more especially in the functions of intellectual expression. Aphasia is therefore a disorder of the symbolizing and categorizing functions, those traditionally referred to as 'superior', and for which pathology seemed to have discovered a localisation, at least to a certain extent. Consequently even for 'superior' functions of the psyche the need for 'reduction' is posited.

The scientific debate in France concerned with the analysis of sensations shows how it is possible to evaluate experimentally the set of problems which had always been thought of as primary evidence for the existence of 'laws of thought', namely the set of problems around language.

These problems were not given particular attention by British anatomic pathology, which followed in this respect the French and above all Broca[43]; they were, however, to receive the attention of Taine in the book we have referred to many times already. Here Taine is guided by sensist ideas influenced moreover by J. St. Mill's theory of the name.[44]

Taine sees the kind of interpretative theories that the 'analysis of sensations' carried out by physiology occasioned as proving the need for any investigation of the mind's activity to be based on physiological data. At the same time he also recognises, in line with what had emerged from British associationism, how inadequate physiological methods were in arriving at a comprehensive investigation of the workings of the "mind".[45]

This consideration means that Taine, although strongly keeping to his belief that 'parallel series' exist between the nerve centres and 'moral events', constantly finds it necessary to refer to introspective analysis.[46] And yet Taine on more than one occasion does not hesitate to reaffirm the superiority of human intelligence over that of animals; and indeed what man possesses is language, the ability to seize upon the common features of several objects and the ability to express them through mime or by means of a 'sign' or symbol.[47]

(6) It is possible then to make a direct comparison between many aspects of French and English psychological and physiological (and indeed psychiatric) thought of the time. The basis for such a comparison should be the two different interpretations of the Cartesian division. A similar comparison with work from German-speaking countries is not, however, possible. From the Thirties to the Sixties they were concentrating on trying to explain the psyche systematically and scientifically, all the way from the analysis of sensations to the genesis of psychic disorders.

During this period German scholars undertook a systematic examination of psychic phenomena. This work helped to define the approach to the study of the 'normal mind', which, as Bynum repeatedly says, British Psychiatry needed in order to gain scientific legitimacy for its 'moral therapy' theory.

The examination conducted by the Germans produced its first significant results in the field of Psychiatry in 1845 with the *Pathologie und Therapie der psychischen Krankheiten* by Wilhelm Griesinger.[48] Griesinger takes up a position on the problem of defining mental illness: he tackles the question of what causes it.[49] Relying considerably on the physiology of Johannes Müller he investigates the relation between the so-called psychic disorders (*Störungen*) and malfunctions, that can be detected by pathological anatomy.[50] Then he goes on to identify in the functions of the '*Vorstellen*' the specific location of psychic activity, marked by tensions (*Strebungen*), impulses (*Triebe*), inhibitions (*Hemmungen*), and repressions (*Verdrängungen*).[51]

Indeed, even before Griesinger, German psychologists (and for that matter French and English too) had shown a general, if not 'technical', interest in the problem of insanity, and, more broadly, in the problem of illness of the psyche. One only has to think of the large number of journals[52] devoted to these problems which appeared in Germany at the end of the Eighteenth and the beginning of the Nineteenth century, or, indeed, of the importance such questions assumed as the Romantic movement got under way. This movement had its final flourishing with C. G. Carus whose theories, as it happens, were on several counts dramatically opposed to those of Griesinger but most of all on the subject of psychic individuality.[53]

All these are facts beyond dispute. But it is equally certain, on the other hand, that Griesinger – much more than other scholars (whom Bynum recalls) closer to romantic or '*naturphilosophisch*' thinking – established an approach to mental illness that was profoundly innovatory. It was so profoundly innovatory because it had so many links, implicit rather than explicit ones, with the progress that German speaking scholars were making towards the foundation of psychology as a science. To be more precise, Griesinger's approach to the problem of mental illness was conditioned by two factors: first, by the process of renewal that had already begun in the Twenties with the psychological analysis of Herbart and Beneke, which was developed up until the Fifties by Lotze and Fechner. And the second factor that provided essential support for the renewal was a line of physiological research which, after Müller, with Ludwig, E. Du Bois-Reymond, and Helmholtz, achieved some important discoveries.

At the same time certain lines of philosophical thought then being pursued had an influence on Griesinger that should not be underrated, orienting and organising his thought. This influence, however, extended to the whole German psychological debate of the time. The ideas that had this marked effect constituted a true 'influential metaphysics' and they were of diverse origins (Leibniz, Kant, Herbart, Schelling, Romanticism). They all shared, though, the characteristic of wanting to settle, in the name of Leibnizian 'pre-established harmony'', the problem of the relation between mind and brain.

(7) The tension between 'laws of thought' and 'analysis of sensation' that steers the development of the debate around the scientific study of the mind is therefore also present, albeit with particular features of its own, within the scientific and philosophical debate in Germany, and one should always keep in mind that this debate started when the methodological ideas of European positivism were in circulation, from the middle 1840's onwards.[54]

Much more so perhaps than in Great Britain and France, the tension between 'laws of thought' and 'analysis of sensations' was significantly distinguished in Germany by the growth of physiology which had started in the Thirties. With their discoveries and sistematising work, great physiologists such as J. Müller, E. H. Weber, E. Du Bois-Reymond, E. Hering, C. Ludwig formulated an analysis of sensations which was destined to reform the categorising framework used in the analysis of the sense-perception system.[55] At the same time this framework came under very considerable pressure from those groups of philosophers who were concerned with developing the Kantian proposal of defining the "conditions of possibility of experience".

Even one takes into account the debt German philosophical thinking owed to British ideas, above all to Mill's theory of induction and unconscious inference,[56] nevertheless the theory that the British led the way in the field of the "scientific study of the mind", although 'formally' correct, consequently needs to be carefully considered.

Indeed it was as a result of the awareness a large part of the scientific and philosophical debate in Germany had of the relationship between scientific investigation and philosophical thinking that gave a decisive impetus to the proposal to establish psychology as a science, as psychophysiology. It was during the period from the middle Fifties to the middle Seventies that this idea was first realised.[57]

The various works that appeared in Germany from 1852 to 1874 were to play a fundamental role in the history of scientific psychology.[58] They outline a way to approach the extent of the workings of the mind, of the psyche, something which had been identified and defined in various ways and 'encircled' by the 'analysis of sensations' and the debate around the 'laws of thought'. This led to Wundt quite explicitly stating the possibility of founding psychology as an autonomous science.

Indeed it is generally true that Wundt's proposal for psychophysiology[59] came about by means of a certain pressurising of the caution that in the face of any theory decidely 'reductionist' or decidely 'spiritualistic' had been present not only in the work of Helmholtz but also in Fechner's psychophysics.[60]

All the same this proposal represented a point of reference for the growth of scientific psychology and was guaranteed by the convergence of the physiological analysis of sensations with philosophical thinking around the 'laws of thought'. At this point it will be helpful to recall the main points of Wundt's proposal for psychophysiology, as he had set out at the beginning of the sixties.[61] They are:

(1) The assessment (under the guidance of J. St. Mill and Helmholtz), as a process of "unconscious inference", of the process of forming perception, which results from the comparison between sensations of relative complexity. These, in their turn, derive from the comparison between 'elementary' sensations, which are unconsciously felt and registered by the sense organs and it is this which can be experimentally observed and evaluated.

(2) The theory that Weber-Fechner's psychophysical law concerning the ratios between stimulus and sensation was significant also from a psychological point of view, or rather was *essentially* psychological in nature.

(3) And finally the theory that scientific experiments in psychology were possible. This opinion, consequent on the previous two points, is based –

and this applies to scientific verification too – on the consistent nature of the laws of psychic processes.

(8) The proposal to establish psychology as a science characterised a large part of the philosophical and scientific debate in Europe from 1830 to around 1870 but it found a more systematically planned approach within the debate that went on in Germany and more specifically in that group of thinkers that were closer – albeit with particular characteristics of their own – to some of the ideas associated with the positivism in the rest of Europe.[62]

This consideration permits us to state quite clearly that the establishment of psychology as a science – we naturally cannot say 'discovery' – resulted from the convergence of a great many lines of scientific inquiry and of philosophical thinking, which had grown up and developed in Europe as a whole. Obviously this development does not end with Wundt's formulation of his proposal but it is still true that generally it represents a unique term of reference if we wish – as it is now necessary to do – to distinguish clearly those features which resulted from the tension, from the early Sixties to the middle Seventies, between the 'analysis of sensations' and the 'laws of thought', a tension that came to represent in essence – but no less clearly for that – those aspects which were to emerge as undeniably problematical in the development of psychology as a science.

This is true, in the first place, for the theory which was central to Wundt's statement concerning the possibility of psychological experiments. This was the theory maintaining that the consistent nature of the laws of physiological processes coincide with the consistency evidenced in the way the 'laws of thought' guide the cognitive process and guarantee logical inferences.[63]

In fact, however much he may believe in the individual character of psychophysiological methodology, Wundt is keen to emphasize that the physical (physiological) level and the psychic level of the sense-perception system (and of the whole cognitive process) are nothing more than two aspects of one and the same dimension.[64] And yet at the very same time he maintains this, he is seen on every occasion to recognise the primacy of the mathematical sciences, which are established and guaranteed by a formal structure. When he examines and attempts to define the features of the 'laws of thought' operating within a logical inference, Wundt indeed seems (and this was more apparent as time went on) not to maintain that the laws of thought are a product of a psychic development of an empirico-associationistic nature. Adopting and in certain cases laying stress on some of Kant's theories present also in the same methodological debate in Britain (above all in Whewell's work), Wundt is thus constantly seen to proceed from the acknowledgment that strong categorising structures exist together with constant and 'deep' logical forms that ensure the growth and organisation of knowledge. Indeed Wundt seems to proceed quite definitely from an acceptance of mathematical conventions as a control to measure the accuracy of any science and model for any logical inference.[65]

But at the same time – and this ambiguity is at the heart of many of the more problematical aspects of Wundt's work – it is also true that Wundt is aware of the need to stress the individual nature of psychic phenomena and of the laws they obey. Here Wundt takes note of what British associationism had learned and in particular J. St. Mill's 'mental chemistry'. This awareness, which was to become very important during the Eighties when the problem of psychic causality was emerging as increasingly significant, was, however, already present in the proposal for a psychophysiology when it was first formulated. It was then further clarified in the first systematic statement of the proposal that appeared in the years 1863–64, especially where Wundt pays particular attention to the relationship between psychological and linguistic analysis. In view of this the influence of Darwinism is not negligible, despite the many problems involved.[66]

More specifically, Wundt examines the question of the origin of language by adopting a two-fold attitude, although his position was gradually to by contaminated by Humboldtian ideas.[67] On the hand, he is essentially alien to any kind of 'reductionism', only concerned as it is with the physiological basis of phonetic laws, while on the other hand he is extremely indifferent to any exaltation of the 'spirit' as an entity that transcends any historical dimension. As Wundt explains in the works where he systematically formulated the new science of psychophysiology, the forms of language are evidence for the 'natural' instrument encountering the categorising apparatus, which results in the production of 'signs', symbols and in the expression of thought. The 'laws of thought' – the 'limits' of our knowledge – cannot consequently escape definition when it comes to the nature of the concrete way they operate in linguistic expression. This is achieved following a historical morphological method which, such as it is, refrains from moving along any 'path into the interior' and at the same time rejects any generic 'reductionism', and which tends to emphasize – significantly in line with British ideas on this point – all the problems around the primary distinction between 'mind' and 'brain'.

(9) In any case, as we have already hinted, even in Great Britain at the beginning of the Sixties, psychologists began to doubt the validity of the distinction between mind and brain. But psychiatrists did not. At the same moment much attention was being given in philosophical circles to the problem of free will and consciousness (already dealt with by J. St. Mill).[68] Another important development at the same time was the growing support, amongst physiologists studying the organism, for the theory that there was a connection between organic processes and mechanical processes, as had been suggested by the principle of conservation of energy in physics.[69]

Besides, for instance, the physiologist Carpenter, there was another man who helped to bring about this fresh approach to the mind-body problem. It was A. Bain. Bain was engaged during the Sixties in working out his hypothesis of psychophysical parallelism.[70] Now, this hypothesis was by no means free of problems. But it does show clearly how important is the reaction of the sense-perception system to the stimuli of pleasure and pain. Something, it is true, that Griesinger (and Müller) had already shown. But without doubt Bain's hypothesis[71] can explain a lot more than the theories of Bucknill and Tuke, who were, as Bynum reminds us, insufficiently resistant to the attractions of "brain mythology".

In fact, with Bain, we are on the threshold of an extremely important decade of the history of British study of the phenomena of the psyche. In 1876 and 1878 we have the foundation of the two periodicals *Mind* and *Brain* – almost a formal declaration that the Cartesian division had become unbridgeable. This decade was important for British Psychiatry too. In the years 1865 to about 1875 it underwent a series of changes, changes which, as Bynum quite clearly states, should be linked with the institutional and social problems caused by the increasing size of the asylums, rather than with increasing importance of neurophysiology.

As we have already pointed out, psychiatrists seem very slow to take account of new ideas. Though, perhaps, they were only following the example of a good number of traditional psychologists.[72] At the same time, however, there were indeed those who declared themselves convinced that, in order to deal with mental illness, even a general practitioner needed to be acquainted with "psychological medicine".[73] But Henry Maudsley was the only one who unreservedly maintained that the results of physiological and psychophysiological research were indispensable for those studying mental illness.

Maudsley was, as Bynum points out, the only English psychiatrist after Cullen to enjoy fame throughout Europe. Without doubt he had something new to say. Maudsley was decidely averse to all forms of introspection –

though in a way completely different from Bucknill and Tuke. He held the problem of mental illness ought to be treated exactly as a problem of 'normal' psychology. Indeed he thought that, if psychology was to be set up as an effective experimental science, then the phenomena of mental illness should serve as an essential point of comparison. As he says:[74] "In reality the phenomena of insanity, presenting a variation of condition which cannot be produced artificially – the *instantia contradictoria* – furnish what (. . .) ought to have been seized with the utmost eagerness; namely, actual experiments well suited to correct false generalizations and to establish the principles of a truly inductive science". In this way Maudsley significantly removes himself from traditional British Psychiatry, with which he seems in fact to have little in common other than his sincere support of non-restraint therapy.[75]

So Maudsley stresses the need for psychiatrists to take note of the developments in psychophysiology. But he attaches great importance too to a number of philosophical arguments which originated from Comte and were, thus, clearly opposed to any non-reductionist conception of psychology. But, above all, Maudsley's discussion abounds, as can be seen at a glance, with arguments of German origin.[76] Such arguments aroused the opposition or at least the suspicion of that large section of British philosophers and psychologists of the Sixties and Seventies who were steeped in the 'science of mind', and whose most authoritative representative in many ways was J. St. Mill. Maudsley and Mill clashed in a bitter exchange. Maudsley accused the author of the *System of Logic* of ignorance in physiological matters. Thus, in the case of Mill, he made several criticisms that he did not need do make against Bain, though he was no nearer to sharing the latter's views on the concept of mind.[77]

(10) However at this point we need to point out the extent to which Maudsley's work was influenced by the problems involved in the "collective foundation" of psychology as a science, which was one of the more complex and productive results of the relationship between scientific inquiry and philosophical thought. Although formulated and developed against the background of European positivism, at the same time however it seemed to put a positivistic 'view of the world' under considerable pressure. This critical state of affairs was reached just before the dispute between 'natural sciences' and 'humanities' erupted. This is not a mere coincidence; indeed one could say that this dispute erupted for the most part precisely because of the many problems encountered in trying to render the 'study of the mind' more

scientific, an attempt which was part of the proposal to establish psychology as a science.

And so it is precisely psychology that seems to question the positivistic model of science, based on the latter's accepted ability of prediction. Can psychology really predict? Can it point to and determine laws of human behaviour? And the question that J. St. Mill had been particular aware of,[78] Can psychology define the effective operating field of human freedom?.

At this point there still remains the possibility of a 'reductionistic' or deterministic solution that was Comtian in character and proposed within the framework of a concept of psychology which, when not definitely in opposition to, was remote from the proposal of scientific psychology developed in the Sixties and Seventies.

This proposal which, was inspired by the tension between 'laws of thought' and 'analysis of sensations', can, it seems, have two possible outcomes:

(1) Psychology is not capable of predicting and consequently proves the 'irreducibility' of the spirit. On the other hand, it considers ultimately the physical-mechanical model of science as the only legitimate one, although still keeping in mind those general developments and increasing number of problems which were to affect this very model from the Eighties onwards.

(2) Psychology tends to structure itself according to different scientific models, associating itself with the biological disciplines and those of a linguistic, philological and historical structure. It is attracted to a model that clearly puts the emphasis on the historical, cultural, and environmental aspects, in an attempt to set up a line of defence against any physical and physiological determinism. It takes up a position therefore that sees as legitimate the genetic-morphological approach but for a whole number of extremely complicated reasons, often different according to national cultures, does not succeed however in adopting and developing, fully and in an effectively original way, the perspectives the evolutionist vision of Darwin had opened up. In fact these two outcomes of the proposal to establish psychology as a science can only be clearly distinguished at the price of greatly simplifying the remarkable complexity of such a programme.

However, granted we can actually talk of a proposal to establish psychology as a science, this is often characterised by these two hardly distinct perspectives overlapping. As long as any attempt to establish psychology has a basis which is in any way limited only to the spirit or only to matter, it is bound to fail to do justice to the complexity of psychic phenomena. In view of this there is a tendency to consider the scientific nature of psychology as being founded and guaranteed by integrating the two possible methodological approaches. And so the belief that gains ground in the course of the Eighties is that on the one hand psychology needs to adopt the genetic-morphological model and on the other hand it needs to adopt the theory of psychophysical parallelism; and indeed this parallelistic model seems to be the most suited to holding together the data and theories that had been developed during the 'collective establishment' of psychology as a science.

(11) All these are basic points to keep in mind when assessing the work of Maudsley as a whole. The special characteristic of Maudsley's approach, however, is, very broadly, the close link he maintains between the physiology and the pathology of mind. With this as a basis he is able to deal with the problem of aetiology of mental illness. This problem was quite clearly spotted by physiologists, and first among them by Carpenter, but was almost completely disregarded, as Bynum points out, by British psychiatrists until the end of the Seventies. But Maudsley wanted to develop his psychopathological views in full awareness of the problem that was preoccupying European psychologists of the Seventies. I mean the problem of psychic causality, which was destined to put in jeopardy the parallelistic model of Lotze and Fechner which Bain had taken over and adapted.

At the same time he did not back away from the problems of 'practical' philosophy which are implicit in discussion of the aetiology of mental illness. I do not have time to go into detail on the ideological and political implicatins of the question. Nevertheless, it must be stressed that Maudsley, although a determinist, was not indifferent to the fact that the mechanisms of heredity involved classifying innocent people as insane.[79]

Maudsley's discussion, then, sets us face to face with many of the most problematical aspects of the psychological debate, and more specifically, of the British psychiatric debate, in the final quarter of the last century. It is really from the point of view of the history of psychiatry that I think the most markedly materialistic and atheistic elements of his thinking should not be underrated. Further we should remember his decision to abandon the editorship of the *Journal of Mental Science*, and, more importantly, the sort of reception given to his works by the majority of British psychologists.[80]

Maudsley was a dogged opponent of the 'metaphysical' conception of mind.[81] In his view – and it is quite clear this position gave rise to the hostility he met in many areas in spite of the respect generally felt for his 'technical' competence – in his view, every study of the workings of the mind ("the most complex and special form of life which we have to do with")[82] had to be founded in the "new truth" that had been revealed by the "discoveries of modern physiology". These discoveries had shown that the consciousness is not co-existent with the mind. The consciousness is nothing other than an "incidental accompaniment of mind".[83] The mind, in turn, is "far from (. . .) being always action", since, in reality, "at each moment the greater part of the mental power exists in statical equilibrium as well as in manifested energy".[84]

Maudsley's arguments can be connected with the theories of Laycock and of Carpenter on 'unconscious cerebration" and are the beginning of what was to become a lively debate.[85] They served Maudsley to demonstrate clearly that physiology would play the decisive role in the study of mental organization. For most of one's psychic life the mind is only potential energy. Clearly then it cannot be studied by means of introspection, by appeal to the consciousness, or by the traditional arguments of the 'science of mind'.[86]

Maudsley, who, however, does not hesitate to reject all panpsychic views, ends up in complete opposition to the theories claiming that psychic life and consciousness coincide.[87] For one thing this set him against a large part of contemporary British culture, but it showed too how much ground separated him from what had been and to a certain extent still was a theory central to the German psychological and psychiatric debate. At the basis of this theory there had been a conception of moral responsability and individual dignity – and of the supremacy of mankind – which it had become more difficult to hold on to before a changing view of society and nature. Indeed, towards the end of the Seventies, optimism and belief in progress gave way, in a large part of European culture, to a more sceptical outlook – an outlook which, though not quite pessimistic, was shaped by the convinction that the individual could not exercise genuine free-will.[88]

Certainly this type of scepticism – with its inspiration in basic determinism – often takes on an extremely questionable ideological colouring – often a racist one. At the same time, however, and this is just an example, we certainly cannot join Stout,[89] who is otherwise a reliable authority, in terming "a stupid prophecy" and "a gratuitous folly" Maudsley's convinction that the human species is condemned to an inevitable decline. But Maudsley believed this because he thought the human species too complicated in its organization to withstand all the future climatical and geological upheavals of the earth.

Sometimes Maudsley does put things harshly. Sometimes he deliberately avoids subtleties. But he does not try to escape problems. It would be difficult not to appreciate his intellectual honesty, the way he supports a scientifically objective view of mental illness and avoids the sort of right-minded philanthropism that had on occasion been used by those wanting to make a case for 'moral therapy". As Maudsley himself says, unyieldingly, "conscious method has not greater part in the formation of moral sense in the later epochs than it probably had in the discovery of fire and its uses in the earlier epochs of human evolution".[90]

Università di Firenze

NOTES

1 On the spreading of Phrenology, see Young, R. M.: 1970; Lanteri-Laura, G.: 1970; Di Giustino, D.: 1975; Cantor, G. N.: 1975[a], 1975[b]; Shapin, S.: 1975; Cooter, R. 1976[a], 1976[b].
2 See Smith, R.: 1973; Daston, L. J.: 1978.
3 Mill, J. St.: (1859) 1978, p. 341.
4 Mill, J. St.: (1859) 1978, pp. 347–348.
5 Mill, J. St.: (1843) 1973–1974, pp. 844–860; 849–851.
6 Mill, J. St.: (1859) 1978, p. 348.
7 Ibidem.
8 Mill, J. St.: (1859) 1978, p. 355.
9 Mill, J. St.: (1843) 1973–1974, p. 852.
10 Bain, A.: 1868, pp. 91, 321.
11 Mill, J. St.: (1843) 1973–1974, p. 8. See also Mill, J. St.: (1865) 1867, pp. 6, 422, 426.
12 Young, R. M.: 1970; Smith, R.: 1973; Hall, V. M. D.: 1979; Daston, L. J.: 1978.
13 See note 1.
14 Mill, J. St.: (1859) 1978, p. 342. On Spencer's psychology, see Young, R. M.: 1970.
15 Spencer, H.: (1855) 1870–1872, vol. II, pp. 291–296.
16 Mill, J. St.: (1843) 1973–1974, Book IV, Chapter II, § 3.
17 See, for instance, Hall, V. M. D : 1979.
18 See, besides Hall, V. M. D : 1979, Cooter, R.: 1976[a].
19 We have in mind, for instance, E. Zola and his strict connection with Bernard's 'milieu'. See also Ribot, Th.: 1873.
20 Ribot, Th.: 1870, pp. 41 ff.
21 Ribot, Th.: 1870, pp. 17–18. The 'inconnu', the 'au delà': pp. 175–230.
22 Ribot, Th.: 1870, pp. 30 ff. Against Comte (and Broussais), p. 22. Against Jouffroy and Maine de Biran, p. 20.
23 Ribot, Th.: 1870, pp. 22, 41, 96–97.
24 Temkin, O.: 1977, pp. 317–339.
25 Ribot, Th.: 1870, pp. 19 ff.
26 Liard, L.: 1878 and 1879.
27 Littré, E.: 1876, pp. 265–266.
28 Taine, H : 1870. On Taine, it is always useful Barzellotti, G.: 1900.
29 Taine, H.: 1870, vol. I, p. 4. On Condillac see vol. II, pp. 265 ff. See also Barzellotti, G.: 1900, pp. 131 and 189 ff.

[30] Taine, H.: 1870, vol. I, p. 361; vol. II, pp. 298 ff. Vol. I, pp. 194 ff.; vol. III, pp. 18 ff.; vol. II, pp. 332, 13, 25–34, 308 ff., 386, 467.
[31] Taine, H.: 1870, vol. I, pp. 16, 22.
[32] Taine, H.: 1870, vol. I, pp. 435–436, 477; vol. II, pp. 68, 160.
[33] Taine, H.: 1870, pp. 367, 70–71, 362–363.
[34] Conry, Y.: 1974.
[35] Taine, H.: 1870, vol. I, pp. 194, 223, 253, 294, 304, 336, 388; vol. II, pp. 76, 110.
[36] Besides Temkin, O.: 1977, see Olmsted, J. M. D.: 1944 and Cranefield, P. F.: 1974. It is always necessary to refer to Flourens, P. M.: 1858.
[37] Hécaen, H.-Dubois, J.: 1969 and Head, H.: 1926.
[38] Lordat, J.: 1843, pp. 141 ff.; Trousseau, A.: 1864, pp. 241 ff.
[39] Broca, P.: 1861, pp. 54–91; Broca, P.: 1865, pp. 267–274; 108–123.
[40] Baillarger, J.: 1865, p. 177; Bouillaud, J. B.: 1825, pp. 29–30; Broca, P.: 1861, p. 62; pp. 63–64, 67; Broca, P.: 1865, pp. 113–117; Lordat, J.: 1843, pp. 130–167; Trousseau, A.: 1864, p. 255, pp. 265–266.
[41] Broca, P.: 1861, pp. 67, 69–70, 72 ff., 89; Bouillaud, J. B.: 1825, pp. 16–17; Trousseau, A.: 1864, pp. 225 ff.
[42] Trousseau, A.: 1864, pp. 265–266; Head, H.: 1926, vol. I, pp. 13–29.
[43] Head, H.: 1926, vol. I, pp. 30–53.
[44] Taine, H.: 1870, vol. I, pp. 472–475. See also Parish, E.: 1897.
[45] Taine, H.: vol. I, pp. 38 ff.; pp. 140–141; vol. II, pp. 293 ff., 300 ff., 332–341, 463–492.
[46] Taine, H.: 1870, vol. I, pp. 140, 277–281.
[47] Taine, H.: 1870, vol. I, pp. 142 ff., 280, 24 ff., 32–38.
[48] Griesinger, W.: 1845; some general references are given by Mette, A.: 1976.
[49] See, for instance, Griesinger, W.: 1845, pp. 99–100.
[50] Griesinger, W.: 1845, pp. 21–22.
[51] Griesinger, W.: 1845, pp. 18–20, 25.
[52] A good general account is given by Dessoir, M.: 1897–1902. See also Leibbrand, W.: 1956, and above all Engelhardt, D. von: 1978.
[53] Carus, C. G.: (1860) 1975, p. 479.
[54] Lange, F. A.: 1873–1875; Helmholtz, H.: (1856–1867) 1909–1911.
[55] Boring, E. G.: 1950.
[56] Helmholtz, H.: (1856–1867) 1909–1911, vol. III, p. 5; Wundt, W.: 1863–1864, vol. I, pp. 58 ff., 131 ff.; Lange, F. A.: 1873–1875, vol. II, Chapter I.
[57] Stanley Hall, G.: 1912.
[58] Lotze, R. H.: 1852; Fechner, G. Th.: 1860, Wundt, W.: 1862; Wundt, W.: 1863–1864; Wundt, W.: 1873–1874.
[59] Wundt, W.: 1862, *Einleitung*.
[60] Fechner, G. Th.: 1882.
[61] Wundt, W.: 1862, pp. XXIV–XXVI; XXVII–XXXII; Wundt, W.: 1863–1864, pp. 85, 86–87, 89–90, 98, 109, 130, 132–137.
[62] Danzinger, K.: 1979; Leary, D. E.: 1979; Bringmann, G. *et al.*: 1980.
[63] Wundt, W.: 1863–1864, vol. I, pp. 198–201.
[64] Wundt, W.: 1880–1883.
[65] Wundt, W.: 1873–1874, pp. 630 ff.; 708 ff.

[66] Wundt, W.; 1863–1864, vol. II, pp. 364–397; 460–461. On these problems see Jankowski, K. R.: 1972; Blumenthal, A. L.: 1970.
[67] Wundt, W.: 1863–1864, vol. II, p. 366 and Wundt, W.: 1901.
[68] Mill, J. St.: (1843) 1973–1974, Book VI, Chapter II, *On Liberty*.
[69] Hall, V. M. D.: 1979.
[70] See, for instance, Bain, A.: 1868, p. 698; Bain, A.: 1873.
[71] See, for instance, Bain, A.: 1868, pp. 300–305; 280 ff. and Bain, A.: 1869, pp. 11 ff.
[72] *Edinburgh Review*, 131, 1870, pp. 437–448.
[73] *Edinburgh Review*, 131, 1870, p. 447.
[74] Maudsley, H.: 1876, p. 21.
[75] Maudsley, H.: 1879, Chapter XI.
[76] See, for instance, Maudsley, H.: 1876, pp. 223, 227 (on Helmholtz), p. 352 (on Herbart and Griesinger).
[77] Maudsley, H.: 1876, pp. 126, 133, 170. Against J. St. Mill, p. 69.
[78] In a letter (3.XI.1843) to Tocqueville (J. St. Mill, *The Earlier Letters of John Stuart Mill 1812–1848*, ed. by F. Mineka, with an Introduction by F. A. Hayek, University of Toronto Press-Routledge & Kegan Paul, Toronto-Buffalo-London, 2 vols. [*Collected Works of John Stuart Mill*, vols. 12–13], vol. II, p. 612) Mill stresses on Chapter II, *On Liberty and Necessity* of Book VI of Mill, J. St.: (1843) 1973–1974: this Chapter is the most important in the book.
[79] Maudsley, H.: 1879, Chapters III and IV.
[80] Croom Robertson, G.: 1877; Stout, G. F.: 1884.
[81] Maudsley, H.: 1876, p. 38.
[82] Maudsley, H.: 1876, p. 43.
[83] Maudsley, H.: 1876, p. 25.
[84] Maudsley, H.: 1876, p. 29.
[85] Hall, V. M. D.: 1979; Carpenter, B.: (1876) 1882; *Edinburgh Review*, 149, 1879, pp. 59–83.
[86] Maudsley, H.: 1876, p. 29.
[87] Maudsley, H.: 1876, pp. 204–205.
[88] Carpenter, B.: (1876) 1882, *Preface*; see also Daston, L. J.: 1978.
[89] Stout, G. F.: 1884, p. 141.
[90] Maudsley, H.: 1876, p. 58.

REFERENCES

Baillarger, J.: 1865, *Recherches sur les maladies mentales*, Masson et Cie, Paris, in: Hécaen, H.-Dubois, J.: 1969.

Bain, A.: 1868, *The Senses and the Intellect*, Longmans-Green & Co., London.

Bain, A.: 1873, *Mind and Body. The Theories of their Relations*, Henry S. King & Co., London.

Barzellotti, G.: 1900, *La philosophie de H. Taine*, Alcan, Paris.

Blumenthal, A. L.: 1970, *Language and Psychology. Historical Aspects of Psycholinguistics*, Wiley and Sons, New York-London-Sidney.

Bouillaud, J. B.: 1825, 'Recherches cliniques propres à démontrer que la perte de la parole correspond à la lésion des lobules antérieurs du cerveau, et à confirmer l'opinion de M. Gall, sur le siège de l'organe du langage articulé', ***Archives générales de Médecine*** **8**, 25–45, in: Hécaen, H.-Dubois, J.: 1969.

Bouillaud, J. B.: 1848, 'Recherches cliniques propres à démontrer que le sens du langage articulé et le principe coordinateur des mouvements de la parole résident dans les lobules antérieurs du cerveau', ***Bulletin de l'Académie royale de Médecine***, 1. er trimestre, 699–719, in: Hécaen, H.-Dubois, J.: 1969.

Boring, E. G.: 1950, ***Sensation and Perception in the History of Experimental Psychology***, Appleton-Century-Crofts, New York.

Bringmann, G.-Tweney, R. D. (eds.): 1980, ***Wundt-Studies***, Hogrefe, Toronto.

Broca, P.: 1861, 'Remarques sur le siège de la faculté du langage articulé, suivies d'une observation d'aphémie (perte de la parole)', ***Bulletin de la Société d'Anthropologie***, 2. ème série, 6, 330–357, in: Hécaen, H.-Dubois, J.: 1969.

Broca, P.: 1865, "Sur le siège de la Faculté du langage articulé', ***Bulletin de la Société d'Anthropologie*** **6**, 337–393.

Cantor, G. N.: 1975a, 'The Edinburgh Phrenology Debate: 1803–1828', ***Annals of Science*** **32**, 219–243.

Cantor, G. N.: 1975b, 'A Critique of Shapin's Social Interpretation of Edinburgh Phrenology Debate', ***Annals of Science*** **32**, 245–256.

Carpenter, B.: (1876) 1882, ***Principles of Mental Physiology***, Appleton & Co., New York.

Carus, C. G.: (1860) 1975, ***Psyche. Zur Entwicklungsgeschichte der Seele. Mit einem Vorwort zur Neuausgabe von Friedrich Arnold***, Wissenschaftliche Buchgesellschaft, Darmstadt.

Cooter, R.: 1976a, 'Phrenology and British Alienists', ***Medical History*** **20**, 1–21; 135–151.

Cooter, R.: 1976b, 'Phrenology: the Provocation of Progress', ***History of Science*** **14**, 211–234.

Conry, Y.: 1974, ***L'introduction du darwinisme en France***, J. Vrin, Paris.

Cranefield, P. F.: 1974, ***The Way In and the Way Out: François Magendie and the Roots of the Spinal Nerves***, Futura, Mount Kisco.

Croom Robertson, G.: 1877, Review of Maudsley's ***Physiology of Mind, Mind*** **2**, 235–238.

Daston, L. J.: 1978, 'British Responses to Psycho-Physiology 1860–1900', ***Isis*** **69**, 192–208.

Danzinger, K.: 1979, 'The Positivist Repudiation of Wundt', ***Journal of the History of the Behavioral Sciences*** **15**, 205–230.

Dessoir, M.: 1897–1902, ***Geschichte der neueren deutschen Psychologie***, Duncker, Berlin, 2 vols.

Di Giustino, D.: 1975, ***Conquest of Mind: Phrenology and Victorian Social Thought***, London.

Engelhardt, D. von: 1978, Bibliographie der Sekundärliteratur zur romantischen Naturforschung und Medizin, in R. Brinkmann (ed.), ***Romantik in Deutschland***, Kohlhammer, Stuttgart, 307–330.

Fechner, G. Th.: 1860, ***Elemente der Psychophysik***, Breitkopf und Härtel, Leipzig, 2 vols.

Fechner, G. Th.: 1882, *Revision der Hauptpunkte der Psychophysik*, Breitkopf und Härtel, Leipzig.

Flourens, P. M.: 1851, *Examen de la phrénologie*, Hachette, Paris.

Flourens, P. M.: 1858, *De la vie et de l'intelligence*, Garnier, Paris.

Griesinger, W.: 1845, *Die Pathologie und Therapie der psychischen Krankheiten*, Krabbe, Stuttgart.

Hall, V. M. D.: 1979, 'The Contribution of the Physiologist, William Benjamin Carpenter (1813–1885), to the Development of the Principles of the Correlation of Forces and the Conservation of Energy', *Medical History* **24**, 129–155.

Head, H.: 1926, *Aphasia and Kindred Disorders of Speech*, University Press, Cambridge, 2 vols.

Hécaen, H.-Dubois, J.: 1969, *La naissance de la neuropsychologie du langage 1825–1865. Textes et documents*, Flammarion, Paris.

Helmholtz, H. (1856–1867) 1909–1911, *Handbuch der physiologischen Optik*, Voss, Hamburg-Leipzig.

Jankowski, K. R.: 1972, *The Neogrammarians*, Mouton, The Hague.

Lange, F. A.: 1873–1875, *Geschichte des Materialismus und Kritik seiner Bedeutung in der Gegenwart*, Baedeker, Iserlohn.

Lanteri-Laura, G.: 1970, *Histoire de la phrénologie, L'homme et son cerveau selon F. J. Gall*, Presses Universitaires de France, Paris.

Leary, D. E.: 1979, 'Wundt and after: Psychology's Shifting Relations with the Natural Sciences, Social Sciences, and Philosophy', *Journal of the History of the Behavioral Sciences* **15**, 231–241.

Leibbrand, W.: 1956, *Die spekulative Medizin der Romantik*, Claassen Verlag, Hamburg.

Liard, L.: 1878, *Les logiciens anglais contemporains*, Baillière, Paris.

Liard, L.: 1879, *La science positive et la métaphysique*, Baillière, Paris.

Littré, E.: 1876, *Fragments de philosophie positive et de sociologie contemporaine*, Aux Bureaux de La Philosophie Positive, Paris.

Lordat, J.: 1843, 'Analyse de la parole pour servir à la théorie de divers cas d'ALALIE et de PARALALIE (de mutisme et d'imperfection de parler) que les Nosologistes ont mal connus', *Journal de la Société de médecine pratique de Montpellier* 7, 333–353; 8, 1–17, in Hécaen, H.-Dubois, J.: 1969.

Lotze, R. H.: 1852, *Medicinische Psychologie oder Physiologie der Seele*, Weidmann, Leipzig.

Maudsley, H.: 1876, *The Physiology of Mind*, Mac Millan & Co., London.

Maudsley, H.: 1879, *The Pathology of Mind*, Mac Millan & Co., London.

Mette, A.: 1976, *Wilhelm Griesinger. Der Begründer der wissenschaftlichen Psychiatrie in Deutschland*, BSB BG. Teubner Verlagsanstalt, Leipzig.

Mill, J. St.: (1843) 1973–1974, *A System of Logic Ratiocinative and Inductive*, ed. by J. M. Robson, with an Introduction by R. F. McRae, University of Toronto Press-Routledge & Kegan Paul, Toronto-Buffalo-London, 2 vols. (*Collected Works of John Stuart Mill*, vols. 8 and 9).

Mill, J. St.: (1859) 1978, *Essays on Philosophy and the Classics*, ed. by J. M. Robson with an Introduction by F. E. Sparshott, University of Toronto Press-Routledge & Kegan Paul, Toronto-Buffalo-London (*Collected Works of John Stuart Mill*, vol. 11).

Mill, J. St.: (1865) 1867, *Examination of Sir William Hamilton's Philosophy*, Longman, London.

Olmsted, J. M. D.: 1945 *François Magendie, Pioneer in Experimental Physiology and Scientific Medicine in XIX Century France*, Schumann's, New York.

Parish, E.: 1897, *Hallucinations and Illusions: a Study of the Fallacies of Perception*, W. Scott, London.

Ribot, Th.: 1870, *La psychologie anglaise contemporaine*, Librairie philosophique de Ladrange, Paris.

Ribot, Th.: 1873, *L'hérédité; étude psychologique sur ses phénomènes, ses lois, ses causes, ses conséquences*, Librairie philosophique de Ladrange, Paris.

Shapin, S.: 1975, 'Phrenological Knowledge and the Social Structures of Early Nineteenth Century Edinburgh', *Annals of Science* **32**, 219–243.

Smith, R.: 1973, 'The Background of Physiological Psychology in Natural Philosophy', *History of Science* **11**, 75–123.

Spencer, H.: (1855), 1870–1872, *The Principles of Psychology*, Williams & Norgate, London, 2 vols.

Stanley Hall, G.: 1912, *Founders of Modern Psychology*, Appleton, New York-London.

Stout, G. F.: 1884, Review of H. Maudsley, *Body and Will, Mind* 9, 135–138.

Taine, H.: 1870, *De l'intelligence*, Hachette, Paris, 2 vols.

Temkin, O.: 1977, *The Double Face of Janus, and Other Essays in the History of Medicine*, The Johns Hopkins University Press, Baltimore-London.

Trousseau, A.: 1864, *Clinique Médicale de l'Hôtel-Dieu de Paris*, Baillière, Paris, in Hécaen, H.-Dubois, J.: 1969.

Wundt, W.: 1862, *Beiträge zur Theorie der Sinneswahrnehmung*, Winter, Leipzig-Heidelberg.

Wundt, W.: 1863–1864, *Vorlesungen über die Menschen-und Thierseele*, Voss, Leipzig, 2 vols.

Wundt, W.: 1873–1874, *Grundzüge der physiologischen Psychologie*, Engelmann, Leipzig.

Wundt, W.: 1880–1883, *Logik. Eine Untersuchung der Prinzipien der Erkenntnis und der Methoden wissenschaftlicher Forschung, I Band: Erkenntnislehre, II Band: Methodenlehre*, Enke, Stuttgart, 2 vols.

Wundt, W.: 1901, *Sprachgeschichte und Sprachpsychologie, mit Rücksicht auf B. Delbrücks 'Grundfragen der Sprachforschung'*, Engelmann, Leipzig.

Young, R. M.: 1970, *Mind, Brain and Adaptation in the Nineteenth Century*, Clarendon Press, Oxford.

NAME INDEX

SUBJECT INDEX

GPSR Compliance

The European Union's (EU) General Product Safety Regulation (GPSR) is a set of rules that requires consumer products to be safe and our obligations to ensure this.

If you have any concerns about our products, you can contact us on ProductSafety@springernature.com

In case Publisher is established outside the EU, the EU authorized representative is:

Springer Nature Customer Service Center GmbH
Europaplatz 3
69115 Heidelberg, Germany

Batch number: 10761588

Printed by Printforce, the Netherlands